Cyborgs of the Ecocene

For Anthony, Anú and Étain

Cyborgs of the Ecocene

What Robots Can Teach Us about Being Human in a Calamitous World

Louise Veling

berghahn
NEW YORK • OXFORD
www.berghahnbooks.com

Published in 2026 by
Berghahn Books
www.berghahnbooks.com

Library of Congress Cataloging-in-Publication Data
A C.I.P. cataloging record is available from the Library of Congress
Library of Congress Cataloging in Publication Control Number: 2026019791

British Library Cataloguing in Publication Data
A catalogue record for this book is available from the British Library

EU GPSR Authorized Representative
LOGOS EUROPE, 9 rue Nicolas Poussin, 17000, LA ROCHELLE, France
Email: Contact@logoseurope.eu

ISBN 978-1-83695-588-7 hardback
ISBN 978-1-83695-591-7 epub
ISBN 978-1-83695-590-0 web pdf

https://doi.org/10.3167/9781836955887

This open access edition has been made available with funding from KU Focus Digital Lives.

Contents

Acknowledgements

This book would not have been possible without the generosity, guidance and encouragement of many people. I am especially grateful to Professor Mark Maguire (Maynooth University) for his support and insight throughout the project. I would also like to express my sincere gratitude to Dr Conor Mc Ginn (Trinity College Dublin), Professor Alan Winfield (Bristol Robotics Lab), Dr Frank Broz (TU Delft) and Professor Illah Nourbakhsh (Carnegie Mellon University) for their openness, generosity, expertise and assistance with my fieldwork. Equally, I wish to thank the many other people who generously shared their experiences and insights through interviews and informal conversations. Finally, I would also like to acknowledge the anonymous reviewers for their thoughtful feedback, valuable suggestions and unique insights.

INTRODUCTION

Cyborgs, Crisis and the Ecocene

They are sitting on chairs. One person is in a wheelchair. They have been eagerly awaiting Stevie's arrival. He rolls in through the door, haltingly, his mechanism whirring away. He is small and shiny and white.

He comes to a standstill in front of the small group of participants, all older people connected with the Irish charitable organization whose focus is on supporting people to age independently at home. Although his wheels are stationary, the rest of him is not. His head continues to move slightly from side to side, his arms make gestures. His big brown eyes are glinting, two eyebrows move up and down. Right now, they are raised, innocently. His mouth is curled upwards in a cartoon smile. This is decidedly not the Terminator. People lean forward in their chairs, their expressions melt into affectionate smiles. They are amused, sympathetic, excited, enchanted. He is more like a child than they were expecting, but also not quite fully working. Not everyone is impressed, though. Mary, at the back, has her arms folded. 'He's not even good looking', she complains. She'd like some colour. Maybe green, she suggests, for Ireland?

'Hello! I am Stevie!' Stevie surprises everyone with his booming English accent. It is an authoritative and confident voice, like an announcer on the BBC, yet also somehow reassuringly automated. Stevie has regained some authority with his commanding introduction, but this group of older Dubliners are on a day out, and they are here to have fun. They want to test him, see how he reacts. 'How are you, my dear?' Paul drawls in an affected tone, 'how is your highness?' Niall calls him 'Steven'. 'Steven, can you open a fridge door and retrieve food?' But the formality doesn't last. Someone shouts, 'Hey Stevie! Where's that fiver you owe me?' Everyone starts to laugh and relax. A few people clap their hands in delight.

At one point, due to a technical hitch, Stevie starts to talk more slowly. Participants respond in kind by also slowing down. 'Are you not too sure?' someone asks, with concern. Stevie laps it up, 'I don't remember, I'm only two years old!' This leads to a delighted 'I'm getting a real reply back!' Far from being put off by these small errors, the participants vociferously reject the suggestion from the robotics lead that it might be frustrating for them. His imperfection appears to strike them as part of his charm, 'part of the personality'.

Everyone's focus is fixed squarely on Stevie, and any attempts that the researchers make to try to talk to them are resolutely ignored. The group have spotted the limitations of Stevie's wheels, and they start to tease him, testing to see how he will answer them. Stevie, can you run? Can you dance? Play football? Niall says, 'Shamrock Rovers [football club] are looking for a scorer!' Stevie, however, is unflappable. His demeanour and his responses appear remarkably dignified, as well as self- and situationally aware. He cannot play sport, he explains logically, because he has wheels, not legs. He also knows what time of year it is. 'Maybe I'll get a pair of legs for Christmas?' he proposes. Everyone is amazed and thrilled by what appear to be his highly advanced social abilities. 'Ah, he's great craic!' Paul sighs, slapping his knees to emphasize his admiration.

Despite having been told at the beginning of the session, they seem to have forgotten, or have chosen to ignore, the engineer behind the desk furiously typing in the appropriate responses on his laptop. Once he types something in, the off-the-shelf text-to-speech translator generates an audio file, which is broadcast through the speakers on the side of Stevie's head. Despite this 'Wizard of Oz' technique, it is not a straightforward case of deception of unsuspecting, vulnerable users. For the participants, while thoroughly enjoying his lifelike capabilities, there is no illusion that he is alive, or human. After gazing at Stevie lovingly, Paul turns his head to the team and announces matter-of-factly, 'Yes, I could use him for a few hours each day, like a TV'. Niall agrees, 'It's great fun to get to know them. To find out what they can do'. Mary, who has somewhat been won over by the interaction, spots an obvious advantage of a robot compared to a human: 'You can turn him off when you get tired of him' (Field notes 25 October 2018).

When I started researching robots in 2015, it would have been difficult, but not impossible, to predict exactly how robots and Artificial Intelligence (AI) would end up being implicated in projects to upturn the existing world order. Today, those in charge of the world's largest technological companies have amassed so much power and wealth that they have become completely untethered from normal processes of accountability. This has been achieved, for the most part, without robots and AI. However, these technologies have provided the overarching narrative and goal for this project and are widely lauded as promising to accelerate it and bring it to its ultimate conclusion. Depending on who you listen to, this might be an inter-galactic universe colonized by super-intelligent digital beings, or a world of increasing precarity, climate catastrophe, 'enshittification'[1] and authoritarianism.

The contemporary prominence of robotics and AI discourses in business, government and the media might lead you to assume that these are settled concepts, or established, proven technologies. This is not the case. Both robots and AI are concepts that were seized upon, in the case of robots, or invented, in the case of AI, purely as marketing tools. In both cases, those using the terms were tapping into the universal human fascination for objects with human likeness or apparent autonomous behaviour. By astutely infusing potentially useful, but often rather dull, technologies with anthropomorphic behaviours or even just narratives of sentience, marketers have managed to attract a seemingly endless amount of funding. This investment has accelerated to such an extent that Big Tech was expected to spend in excess of $320 billion on AI in 2025 (Morris and Uddin 2025).

Although many different approaches have been used to try to achieve AI, the current AI 'boom' is based on machine- or deep-learning techniques, which use algorithms to interrogate large datasets, looking for patterns. These patterns can then be used to create a predictive model, and, using both the patterns and the content of the datasets, generate AI output, or 'GenAI'. This has, famously, led to the creation of a number of sophisticated chatbots and image/video generators, which have proven highly effective in selling AI's potential as a transformative technology to the world. In order to ensure that this gift keeps on giving, industry leaders are now, once again, promising that they are on a path to imminent human-like intelligence, currently known as Artificial 'General' Intelligence, or AGI.

The term 'artificial intelligence' first appeared in the 1950s in a research proposal led by John McCarthy, an assistant professor of mathematics at Dartmouth College. The name was deliberately chosen to differentiate it from the rival research programmes of cybernetics and behaviourism. In the proposal, the authors claimed that they would 'solve' intelligence in a matter of weeks. They did this primarily, as McCarthy later admitted, to secure funding. Over the following decades, researchers in AI made great strides in highly specific areas, such as proving mathematical theorems, but never managed to develop anything remotely comparable to human intelligence. This led researchers to designate human-like intelligence as 'general': something that is not just successful at achieving a specific, narrowly defined task, but could be applied generally to a range of varied tasks, like humans do. The difference between 'narrow' intelligence and 'general', then, is the difference between a chatbot and a human. The latest 'predictions' for achieving AGI by tech leaders vary between one and ten years, with just a few voices, such as Meta's Yann LeCun, arguing that current techniques will not lead to AGI (Kim 2025).

The business case for AI and robotics systems, however, is not about the scientific and philosophical breakthrough of creating human-like intelligence; rather, it is about their ability to replace labour. Recent definitions of AGI have thus switched from a focus on matching human cognitive abilities to a much narrower 'highly autonomous systems that outperform humans at most economically valuable work' (OpenAI 2025). But, profitable as replacing labour in vast swathes of the labour market would be, there is another side to AI that is not just about monetary gain. Since the 1980s, a parallel ideology has emerged. It is based on the transhumanist idea that, through technology, humans can be liberated from their biological constraints within a technological utopia. As such, AI and robotics systems are promised as much more than innovative technical products. As well as the potential to become arbiters of policy, justice, warfare, communication, education, culture, surveillance and control, they are, in this view, also the path to salvation. According to this logic, anything that stands in the way of its rapid and unregulated development, such as AI governance or concerns over the climate impact of its energy demands, is morally corrupt. This is the backdrop to PayPal founder and tech investor Peter Thiel's branding climate and justice campaigner Greta Thunberg and AI 'doomer' Eliezer Yudkowsky as possible candidates for the 'antichrist' (Mak 2025).

Most of the roboticists that I interacted with over my four years of fieldwork firmly distanced themselves from these fantastical and exaggerated narratives of AI and robot capabilities. Instead, roboticists overwhelmingly consider their work to be solving immediate and pragmatic problems. They tend to think of themselves as inventors, creators and tool-makers. Roboticists want to build something new: to 'get things to work'. While many sat behind screens writing code, many also wrestled with spanners and screwdrivers, engaged in such varied activities as connecting electrical circuits, fixing wheels, soldering metal and moulding arms. As a result, many robotics laboratories are more likely to resemble workshops than scientific laboratories, cluttered with bits of machines, robot parts and tools. If you were to ask a roboticist, and

I did, many times, what a robot is, they tended to define it in very technical terms. Most said something along the lines of: a robot is a system that takes an input, processes it somehow and then performs an action. A robot is thus an AI system, with one major difference, which is that it can perform a physical action in the world.

However, despite the insistence from most within the roboticist community that their work is purely technical and rational, the narrow boundaries that they imposed on defining their object would inevitably break down in other contexts. Indeed, robots did not originate in engineering, or business, but in fiction. The original robots first appeared in a 1920 play by Czech playwright Karel Čapek called *R.U.R. (Rossum's Universal Robots)*. Today, the popular conception of robots continues to be shaped by fiction, particularly film, whether it is the 'good' robots, like C-3PO and R2-D2 from *Star Wars* or the 'bad' ones from *The Terminator* and HAL from *2001: A Space Odyssey*. Additionally, as I describe later in this book, despite distancing themselves from it, roboticists are nonetheless implicated in subtly encouraging fantastical discourses about imminent intelligence in order to promote their field and secure research funding. Although most roboticists insist that their work is purely a technical act, others also recognize it as experimental philosophy. For most, however, whichever way they explain it, what they are doing is trying to build a human.

Building Humans

This is a book about robots. More specifically, it is about our attempts to replicate ourselves mechanically, and, as such, it is also a book about being human. The way we conceive of ourselves and our relationship with the world matters. It informs how we organize ourselves and how we act in the world. Although the majority of roboticists claim not to be concerned with philosophical questions, by building humans they are, sometimes explicitly, but more often implicitly, reflecting, reproducing, producing, and sometimes challenging, core assumptions about what it means to be human.

At this particular juncture in history, we are caught between two extreme views of the human. On the one hand, we have the project of anthropocentric humanism. This dominant view of the human today arose in the Enlightenment and was further shaped by neoclassical economic theory. In this view, the human is conceived as an atomistic agent: all-knowing, self-interested, competitive, separated from and dominant over nature. From this perspective, our relationship with nature, and to a large part with others, is exploitative and extractive. And yet, this dominant economic theory is, as Kate Raworth (2022: 6) puts it, 'the mother tongue of public policy, the language of public life, and the mindset that shapes society'. It is this also the implicit paradigm within which much robotics research takes place.

As a counter to this view of the human, contemporary scholarship in Anthropology and Science and Technology Studies (STS) takes a posthuman approach. This challenges human exceptionalism and instead reveals ways in which rigid boundaries between humans and non-humans are unstable and subject to dissolution. Increasing focus on the climate crisis and the emergence of the Anthropocene[2] has made arguments for the dissolution of stable boundaries, including between humans and nature, ever more pressing. For theorists in these fields, such as Donna Haraway (1991), advances in AI and robotics research are proposed as evidence for the increasing obsolescence of the human as a category of ontological significance.

My observations of robots and robotics research, however, do not support either of these views. Instead, as will be discussed throughout this book, both affective encounters with robots and observations of efforts to animate machines reveal the reality behind the hype: the deep asymmetry between human vitality and intelligence, on the one hand, and the lifelessness and inertness of machines when not animated by countless human interventions, on the other. However, far from being a call in favour of human exceptionalism, I argue instead that what is revealed is the irreducibility of embodied life, and a continuity with all other living beings. In the concrete encounter, the living body emerges as a durable category that cannot be reasoned away. In its holistic, embodied sense, it has so far eluded attempts at artificiality and mechanization. Instead, what is revealed is the complexity and ambiguity not just of human sentience and intelligence, but of all living beings, which continues to resist translation into a problem of coding, instrumental control and market exchange. As we will explore, this also reflects emerging scientific research which investigates how our experiencing bodies continue to share a greater commonality with other embodied, living beings, than with machines.

Ultimately, this book argues that to reimagine the human is not to merge it with the machine, but to resituate it within the living. It means to confront and accept asymmetries between us and machines, while also recognizing a radical continuity between humans and other living beings. I will argue that this is an ontological ethic to redefine our relation with the world and with others in a way that will allow for ecological renewal. In order to develop this admittedly ambitious task, I bring ethnographic and empirical material into dialogue with two philosophical orientations that may seem incompatible. The first is Donna Haraway's (1987) Cyborg, a posthuman figuration that challenges boundaries and opens spaces for political reimagination but, I argue, risks collapsing the distinction between the machinic and the organic. The second is the 'proto-ecological' embodied phenomenology of Maurice Merleau-Ponty (1968, 2012) and the 'ecological existentialism' of Deborah Bird Rose (2011), both of which ground the human in embodied, affective, lived experience that emphasizes continuity with the more-than-human world.

Combined with Merleau-Ponty's concepts of embodiment and flesh, Rose's 'kinship of becoming' challenges anthropocentrism without collapsing into machine equivalence. Rather than a generic category of 'non-humans', it is all life forms in their infinite variety, as well as the conditions that make life habitable, that are given moral consideration. This involves clearly demarcating life and ecological systems from technical and commodified objects. Distinguishing living beings, particularly human and animal life, from technical objects and commodities defies the economic framing that reduces life to its instrumental or exchange value. Most importantly, it allows for a recognition of the common vulnerabilities and fates of those of us who share the planet's finite resources and are dependent on its precarious ecological balance for survival.

The concept of the Anthropocene, despite providing some discursive utility, is nonetheless an apolitical diagnosis of the problem. It constructs a singular universal collective humanity, which elides deep inequalities and deflects attention from understanding how such inequalities are generated and intensified. The alternative 'ecocene' is proposed to reorient towards the positive action necessary to live together within a healthy biosphere (Armstrong 2016; Küpers 2020; Tănăsescu 2022).

Empirical and Theoretical Approach

AI and robotics research offer an extreme example through which to investigate our entanglements with technology, and with the world. On the one hand, robots have much in common with other technologies. They have the potential to extend our reach into the world, and to increase our comfort and convenience. As is the case for other technologies, robots are deeply interwoven with economics and politics, and may be used to concentrate power and to shift it (Winner 1980). They are shown to have a strong 'bio-power' effect, affecting bodies and immersing them in social relations of power, inclusion and exclusion (Foucault 2010). Like other technologies, concerns have been raised about the specifics of their technological capability, energy consumption and design. These range from immediate concerns, such as issues of deception and attachment with potentially vulnerable robot users, to job displacement, surveillance and autonomous weapons, and longer-term existential concerns. Thus, as is true for other technologies, the ongoing proliferation of AI and robotics technologies in our lives may represent both an opportunity for the enhancement of life, as well as a tool to negatively affect people's lives, including increasing inequality and suffering in society.

But humanoid robots also occupy a unique position in the contemporary. They are a powerful symbol of advanced technology and human ingenuity. In literature and popular culture, they provide complex mirrors of our culture(s), collective identities and existential fears. The potency of their symbolic power is also connected to their physical nature, and the strong reactions that are evoked in the concrete encounter by their shape and movement in space. No other technology is used so overtly to replicate, and thus understand, ourselves as humans in the world. In a robotics laboratory, core ontological and epistemological assumptions are enacted. As Lucy Suchman (2007: 226) observes, robots act as a 'doubling or mimicry in the machine', a concrete revelatory site that both discloses and produces assumptions about the human, including what is considered essential and what is considered 'other', as well as assumptions about gender, class and race. By building and modifying robots, robotics researchers are thus reflecting, reproducing, producing, and sometimes challenging, core assumptions about what it means to be human.

Robots are at once material and symbolic, scientific and commercial, imaginary, utopian and dystopian. But it is in their resolute materiality that robots differ most profoundly from abstract ideas about AI. AI exists in the digital realm: it is written in code and algorithms, hierarchies and layers, impenetrable to most people and mysterious for that reason. Robots, by contrast, are resolutely material. They are necessarily tethered to the local, and subject to physical laws and contingent contexts. The most promising and profound theories of intelligence have stuttered to a complete standstill, literally, when implemented in a physical system.

Robots embody many contradictions and multiplicities, just as the work of robotics spans multiple territories. Robots are at once physical and local entities, but also part of a wider, global, hyperreality of social connectivities and digital networks. They are experienced at a personal and visceral level but are also part of wider social, cultural and 'socio-technical' imaginaries (Jasanoff and Kim 2015). Robotics researchers are spread across many sites and engaged in many different types of work, including designing and installing physical components, programming behaviours, holding meetings, engaging in demos and networking, writing papers, attending conferences and testing the robots in community settings. In order to fully grasp the complex world of robots and robotics research, I needed to go to the places where robots are conceived, built and tested.

This book draws on ethnographic fieldwork and interviews I conducted between 2016 and 2019 at key robotics research sites in Ireland, the United Kingdom and the United States. I also travelled to Tsukuba, Japan for a robotics conference.[3] Although my background is not overtly technical, I had spent the decade before starting this research project working alongside technical teams both in industry and in academia. Despite this, robotics and AI research are spaces where technical expertise, institutional authority and professional status carry significant weight. Compared to my informants, I had significantly less of these forms of power. This imbalance shaped my approach to the research, requiring me to quickly understand and navigate implicit hierarchies, performing deference, and to adopt the stance of learner rather than investigator. Nonetheless, for the most part my informants welcomed me into this space in a surprisingly open way. I found that many valued alternative perspectives and disciplines because they were aware that the success of their robots required more than just technical expertise. My background in qualitative research and user experience was also considered useful to those within the field, allowing me to participate in robotics studies as part of the robotics team. Despite having some insights into the field, I considered myself an outsider and approached the field with certain preconceptions about engineers and roboticists. However, extended engagement with the people in the field, as well as ongoing reflexivity, particularly during the writing process, allowed me to confront these assumptions and develop more nuanced insights.

Although my research was multi-sited, my primary research site was a robotics laboratory in Trinity College Dublin in Ireland, with which I was engaged on a weekly basis for two years from 2017 to 2019. I first met Daniel,[4] the creator of the robot, Stevie, and head of the team, at the International Conference on Social Robotics (ICSR) in Tsukuba Japan in November 2017. Stevie had been entered in to, and won, one of the prizes in a robot design competition. After returning to Ireland, I started to visit the lab at Trinity regularly over the next two years and it became my primary field site and one of the rare labs that worked on building a complete humanoid robot. I also accompanied the team to places where Stevie was being trialled and tested, including a number of focus groups with older people in Dublin and a retirement community on the east coast of the United States.

As well as the lab at Trinity, I travelled to other labs: the robotics lab at Heriot-Watt University and Bristol Robotics Lab in the UK and the Robotics Institute at Carnegie Mellon University in the US. This multi-sited ethnographic approach (Marcus 1995) allowed me to trace connections between local sites of observation and to start to develop an understanding of the cultural production of robots across a number of localities and temporalities, although primarily within an anglophone context. In this way, I could map associations and connections and pay attention to the continuities and discontinuities between and across sites. This revealed communities of roboticists that are geographically and culturally specific, while also sharing a common identity with the wider robotics communities. Although my engagement with robotics communities outside of these contexts was limited, ethnographic insights from Japanese robotics reveal similar continuities and discontinuities (Bender 2025; Wright 2023).

In all research sites, I observed the work and generally hung out and chatted with researchers in the lab. When I could, I would accompany them to meetings, as well as demonstrations with the media and potential funders, and trials with potential users. In user trials, I generally took a more participative approach to observation, sometimes conducting studies with participants and sometimes becoming the subject of experiments with robots myself. My field research ended just before the COVID-19 pandemic, which had a profound impact on people's perception of the need and acceptance of robots, particularly within care environments (Kodate et al. 2024). However, as I discuss in Chapter 3, people's responses remain remarkably consistent.

Structure

In this book, I bring together ethnographic description, interpretive engagement with existing scholarship and philosophical inquiry to trace robotics research as a practice embedded in specific times, places and relations. Building on this work, I argue that we must recognize an asymmetry between humans and technical objects, while also acknowledging a radical continuity between humans and other living beings. Chapter 1 grounds this narrative in its historical and theoretical context. It explores the key historical events that led to the emergence of robots as they are today, revealing how the human and machine have been co-constructed, with each serving as a model for the other. It shows how robots are at the centre of centuries of attempts by mathematicians, philosophers and engineers to distil the essence of life and humanity. From a focus on those who make these machines, the chapter then moves to anthropological reflections on humans and machines, including Haraway's Cyborg figure.

Chapter 2 looks at the spaces where robots are conceived, built, modified and studied by bringing ethnographic attention to robotics labs and conferences in the UK, Ireland, Japan and the US. I explore both the geographical specificity of individual robotics labs and also a community of researchers united by common histories, identities and mythologies. In this chapter, I introduce Stevie the Robot, a humanoid social and care robot whose development I followed for a period of two years from construction in the lab to the sites where it was trialled and tested. The chapter closes with a discussion of the 'conceptual' space of roboticists' imaginaries, including the materiality and precision of their work and how this influences their perception of the world.

In Chapter 3, the focus moves from the laboratory to the places where robots are tested and interact with potential 'users'. This chapter brings ethnographic attention to specific and situated encounters with robots, in particular illuminating people's intense visceral responses to them. As well as face-to-face encounters, this chapter also focuses on those who work on, and through, the robot. I address the issue of theorizing the encounter in robotics work and instead propose a phenomenological interpretation in contrast with typical metacognitive explanations. I argue that the way in which the robot is experienced in the interpersonal encounter is an embodied experience, revealing the human open to a creative and intersubjective world.

Chapter 4 explores the wider political economy of robotics work. First, I discuss the current and proposed uses of AI and robotics to disrupt the labour market as articulated by the robotics community. This is followed by a discussion of current discourses around machine learning and generative AI. The chapter then moves to elaborating the institutional investment in constructing, promoting and selling a specific hyper-inflated vision of machine intelligence and robot animacy, driven by both the profit imperative and quasi-religious transhumanist utopian visions of technological futures. Next, the chapter discusses the vast and unseen human work that is necessary to animate robots and other 'intelligent' systems, often done by marginalized workers, revealing some of the social and environmental costs of maintaining the illusions of animacy and imminent robot autonomy. The narrative then turns to ways in which, in micro-encounters, this illusion of animacy can also be a social and creative act, revealing how the robot may be used both to concentrate power and to shift it, on both local and global stages.

In Chapter 5, I build on the insights from the previous chapters, as well as scientific research from other fields, to argue both for an asymmetry between humans and machines and for a radical continuity between humans and other embodied, living beings. I reveal how, contrary to the dominant narrative, AI and robotics research reveal a limit to knowledge and the persistent ambiguity of the figure of the human. I discuss the issue of embodiment in robotics work, and in critiquing the work discuss the implications of taking embodiment seriously. I then propose a novel figure of the human by bringing Haraway's Cyborg and the eco-existential human of Merleau-Ponty and Rose into dialectical relation. The book concludes with some of the implications of recognizing an asymmetry between humans and machines, alongside a radical continuity between humans and other living beings.

Notes

1. Coined by Cory Doctorow, 'enshittification' describes the increasing degradation and lowering in quality of the online experience resulting from the profit motive under capitalism.
2. Some have proposed that we refer to our current epoch as the 'Anthropocene' in recognition of the outsized and catastrophic impact that humans are having on the planet. While the Anthropocene has not been accepted as a scientific term, and a proposal to designate it an official scientific epoch or geological event was defeated in 2024, it nonetheless plays a significant discursive role in describing our collective responsibility for the climate crisis and the negative effects of industrialization and unbridled capitalism.
3. My fieldwork was supplemented by more than one hundred informal, open-ended, unstructured interviews, as well as sixteen formal, semi-structured interviews with roboticists. Along with field notes, all interviews were transcribed and transcripts were anonymized. The research underwent ethical review and was granted approval through the university's ethics process.
4. Throughout this book, I use pseudonyms for all my informants, for consistency and to protect their anonymity.

CHAPTER 1

Human–Machine Co-Constitution

A Historical and Philosophical Prelude

By the time I arrive at the lab in Trinity, Stevie the Robot is already two years old. Yet, in his[1] short time as a social and care robot, he has already become an international superstar. Stevie is one of a handful of robots, including R2-D2 and the medical robot Da Vinci, to have appeared on the cover of *TIME* magazine (see Purtill 2019). He is also listed as one of the top one hundred inventions of 2019. His appearance is humanoid but not at all human. He looks like a child robot, or a robot that a child would draw. His voice, on the other hand, is that of an adult male. His accent is English.

Stevie is a historical object. He did not spring fully formed from the imaginations of his creators, nor is he the product of a systematic and methodological application of science. The events that led to Stevie's creation were neither inevitable, nor predictable. Instead, the facts of Stevie's existence: his appearance, the way he is constructed and the technologies that are used to implement his capabilities are the result of specific historical events and social conditions.

The Age of the Automaton

Robots have a lineage stretching back far into pre-history of human attempts to represent and reproduce life and human likeness; for aesthetic purposes, for entertainment, to inspire religious practice and in the pursuit of craft, knowledge and understanding. However, human-likeness as 'technology' is a more recent development.

The view of technology as a symbol of advancement, and the progress of civilization, dates back to Classical Greece. In her study of luck and ethics in Greek tragedy and philosophy, the philosopher Martha Nussbaum (2001) shows how themes of reducing risk or 'exposure to luck' dominated Greek philosophy. Plato's Protagoras paints a picture of humans in nature as isolated, exposed and miserable. Prometheus, in his kindness, grants these pathetic creatures the gift of the *technai*:

> House-building, farming, yoking and taming, metal-working, ship-building, hunting; prophecy, dream-divination, weather-prediction, counting and calculating; articulate speech and writing; the practice of medicine; the art of building dwelling-places ... (2001: 9)

Luck, or *tuchē*, is contrasted with *techne*. The former, according to Democritus, is associated with witlessness, an excuse people give for their own lack of resourcefulness. *Techne*, on the other hand, is a 'deliberate application of human intelligence to some part of the world', making human existence safer, more predictable and allowing control over contingency (Nussbaum 2001: 95).

Automata, or self-operating machines, seemed the ultimate realization of technical advancement. An instrument that could 'accomplish its own work, obeying or anticipating the will of others', wrote Aristotle in 320 BCE, could even replace the need for that pre-eminent and indispensable instrument, the slave (Aristotle 2000: 7). However, it is Hero of Alexandria (10 BCE–70 BCE) who is credited with building the first known automaton: a self-powered three-wheeled cart (Sharkey 2007). In the following years, automata emerged with mechanisms powered by steam and water, cogged gears and pulleys.

Throughout the Middle Ages, efforts at both imagining and building artificial life continued in varying and largely isolated ways until the early Renaissance period. Automata were popularized as demonstrable feats of engineering and human ingenuity, but their ability to inspire awe and wonder in those who encountered them led them also to be used for religious purposes, as well as for entertainment. Sculptor Elizabeth King (2007) writes about her experience with a small automaton in the image of a monk dating from around 1560, which, she theorizes, may have been constructed by engineer Juanelo Turrino for King Philip II of Spain. She describes her own encounter with the small moving object, as well as the experiences of others who encounter it, as 'intimidating', invoking a 'primal anxiety': 'when this machine heads in my direction on a table, my animal flight urge stirs' (King 2007: 274, 277). This response leads her to imagine what it must have felt like for those who encountered the automaton at the time: 'Could it even momentarily have been perceived as alive, just for the space of a shudder?' In seventeenth-century Japan, *karakuri ningyo* puppets were animated using the same mechanism of gears and springs that is used in clocks. The Japanese word 'karakuri' refers to the 'subtle trickery' of devices that evoke a sense of awe and wonder by having their inner workings concealed to the viewer (Shea 2015).

It is little wonder, then, that these displays, along with the emerging new sciences, shaped contemporary conceptions of the human. If automata, with their simple mechanisms, could demonstrate lifelike behaviour, it seemed likely that all life could be reproduced with just slightly more complex mechanisms. In the fourth century BCE, Plato first postulated the idea of mathematics and the ideal world of geometrical forms as the real and true form of reality, while the material world of daily life is merely its shadow (Riskin 2007). However, with the rise of automata, this idea really took hold. These developments led thinkers such as René Descartes to hypothesize the universe, and all it contains, as mechanical. The artificial replication of human-like behaviour and intelligence seemed imminent. The image of the human was thus co-emergent with machines, exemplified by the figure of the automaton.

In her history of AI, Pamela McCorduck (2004) relates how, in 1673, philosopher and mathematician Gottfried Wilhelm Leibniz set about developing an unambiguous formal language to prove that all human activity could be reduced to mathematical calculation. While he did not succeed in this task, he did manage to invent the binary system, and a digital mechanical calculator composed of symbols, called the 'Step Reckoner', a key development in the later creation of the computer. Despite its failure to replicate 'intelligent' behaviour, Leibniz theorized that it might instead be capable of doing the slave-like work of calculation that is 'unworthy of excellent men' (McCorduck 2004: 26). Throughout the eighteenth century, philosophers Spinoza, Hobbes, Locke, Kant and Hume, and scientists La Mettrie and Hartley, all tried and failed to formulate laws of human thought. A proliferation of automata accompanied these scientific attempts.

For historian Jessica Riskin (2003), the point of origin for our contemporary way of thinking about automata is French inventor Jacques de Vaucanson's 'Defecating Duck', first displayed in Paris in 1738. Unlike previous automata, the mechanical Duck was not simply designed to amuse but was also an exercise in experimental philosophy, testing which aspects of living creatures could be reproduced mechanically, and to what degree. Vaucanson's Duck appeared to ingest food and excrete it in an altered form. The Duck was simultaneously a scientific, philosophical and commercial venture. Between 1770 and 1854, another famous automaton, the chess-playing 'Mechanical Turk', was exhibited across Europe and the Americas. Ultimately, however, both Vaucanson's Duck and the Mechanical Turk were revealed as frauds. In the artificial duck, the food and the resulting excretion were in fact completely different substances. Similarly, the Mechanical Turk's mechanism was shown to consist not of a highly sophisticated mechanized human, but of an actual human, hidden underneath, operating the machinery.

Nonetheless, throughout the eighteenth and nineteenth centuries, increasingly intricate and ingenious automata that could write messages, play musical instruments and repeat phrases were built by engineers and clockmakers. There was, however, also a darker side to these figures. Instead of simply representing the genius of mechanical invention, these automata were often also represented as 'exotic' or uncanny characters, often female or black, as animals, acrobats or magicians (Norton-Wise 2007). They were a concrete and physical representation of a particular view of humanity, a mirror against which to explore and produce identity and difference. There was also an emerging discomfort with the new technologies. Mary Shelley's (1993) *Frankenstein*, which was first published in 1818, recounts a tale of horror in which a scientist creates a human-like, yet monstrous, creature by stitching together the remains of corpses and reanimating them using electricity.

By the late seventeenth and early eighteenth centuries, the creation of machines that could emulate life stalled, and shallow, human-like imitations had been relegated to the realm of entertainment and curiosity. At the same time, however, the project of replacing human workers with machines went into overdrive. The 1760s saw the beginning of the Industrial Revolution, with increasing automation, particularly in the cotton and iron industries. The spinning-mule or 'Spinning Jenny' was invented in 1779 to automate the previously manual job of spinning cotton. Now, a greater variety of textiles could be produced more cheaply to many more people, with significant profits for factory owners. This also radically transformed the conditions of the workers, and their relation to their work and its products. These developments inspired novel, and disparate, responses from theorists. Friedrich Engels' father owned a cotton-spinning business based in both Germany and England, and he had seen the effects on workers first hand. This would lead him, along with Karl Marx, to document the resulting unemployment, suppression of wages and deskilling (Boyer 1998).

An entirely different response came from engineer Charles Babbage. For Babbage, observations of the workers at their machines on the factory floor suggested a natural 'unequal division among machines', in which those at the top produce power and those at the bottom produce mechanism, existing 'merely to transmit force and execute work' (Babbage 1832: 16). Babbage developed this insight into a universal classification system that he believed applied universally to machines, to social organization and to individual humans. As Leibniz and others had previously attempted, Babbage hoped to formulate this theory into a universal language. He developed a machine that he called 'the Analytical Engine', which, as well as carrying out calculations, would be 'capable of analysis and tabulating any function whatever' (McCorduck 2004: 31). Although the engine was never built, he had established the principle of what would become the modern digital computer (Dreyfus 1992).

The Rise of the Robot

The world's first 'robot' appeared in a 1922 play *R.U.R. (Rossum's Universal Robots)* by Czech playwright Karel Čapek. The word 'robot' comes from the Czech word *robota*, meaning slave, work or drudgery. In the play, artificial people are created using organic matter so that they might produce goods at a fraction of the cost of human labour. The 'robots' eventually rise up against their creators and overthrow humanity. The play travelled internationally, including in New York in 1922 and in Tokyo, Japan in 1924 under the title *Jinzō Ningen* (Artificial Human). Despite its ambivalent message, it captured something in the public imagination, sparking an enduring fascination with robots.

During these early post-war years, science fiction as a distinct genre also took off, coined by publisher Hugo Gernsback in the first issue of the 1926 science fiction magazine *Amazing Stories*. 'By "scientifiction"', Gernsback writes, 'I mean the Jules Verne, H.G. Wells and Edgar Allan Poe type of story – a charming romance intermingled with scientific fact and prophetic vision' (Westfahl 1992). The following year, Fritz Lang's feature-length film *Metropolis* was released, combining a dystopian, expressionist view of the oppressive and destructive potential of technology with a utopian 'technology cult of the Neue Sachlichkeit and its unbridled confidence in technical progress and social engineering' of the Weimar republic (Huyssen 1986: 223). The film features a stark division between an above-ground futuristic technological utopia, inhabited by the city's elite, and a below-ground dystopia, featuring dark chambers and factories inhabited by the city's workers. The film also portrayed a new image of the robot, or *Machinenmench* (Machine person), an evil female android, Maria.

The same year, the Westinghouse Electric Corporation in Pittsburgh unveiled the 'world's first' humanoid robot, Herbert Televox. The robot owed its existence to an unlikely confluence of events. In the mid-1920s, one of the engineers at Westinghouse, Roy James Wensley, designed a device for changing switches remotely using sound. At one end, a tuning-fork oscillator created a particular frequency which then generated a code. The code was sent via telephone line to a receiver unit, which processed the code and opened or closed a particular switch in response. Wensley called his device the 'Televox' (Marsh 2018). In a review of the device, the *New York Times* described it as an 'electrical man', a mechanical robot or slave that 'obeys without the usual human arguing, impudence or procrastination' (Kaempffert 1927). Although in the main body of the article the author takes pains to distance Wensley's machine from an actual robot, which would be a 'fantastic creation' for which 'the modern engineer has no patience', a new bar had been set.

When Wensley was asked to present his creation at the opening of the Level Club in New York City, he knew that people were expecting more than a 'little frame of relays and wires' (Letizia 2013). Taking his cue from the *New York Times* article, Wensley decided to create a humanoid casing for his machine and give it a name, Herbert. While its torso comprised the Televox control box, it was set into a wooden frame with a head, arms and legs. The 'robot' could accept a telephone call by lifting a receiver, and then 'act' depending on which signal was sent. It also made noises, specifically buzzes and clicks, and waved its arms. Herbert Televox was an international sensation. In a manner that would prefigure the dynamic between robotics researchers, the media and funders for the coming century, Wensley benefitted from the often vastly exaggerated reports of his robot's abilities in the media by attracting funding, while also spending much of his subsequent career trying to correct and downplay the exaggerations (Marsh 2018).

Other robots soon followed. In 1928, a Japanese robot named Gakutensoku, or 'Learning from the Laws of Nature', was created by biologist Makoto Nishimura. Gakutensoku was a very different kind of robot to those being built in Europe and the US. First of all, it was enormous, more than three metres tall. It had a golden upper body and was dressed in a toga. The robot was seated on an altar, resembling a divinity more than a feat of science (Hays 2013). It was powered by a 'ventricle system' consisting of rubber tubes and air compressors, allowing it to change its facial expressions, puff its cheeks, move its chest and simulate writing. For Nishimura, the robot was a part of nature, and he is quoted as saying 'if one considers humans as children of nature, artificial humans created by the hand of man are thus nature's grandchildren' (Whelan 2011). Unlike the portrayal of robots in the US and Europe, Nishimura is reported as insisting that his robot was not a 'slave': '[i]t would be sad to find joy in making a slave-like android that simply copies humans, which are masterpieces of the earth' (Hays 2013).

After Gakutensoku, however, robotics in Japan stalled until the years following the Second World War. After the war, Japan was occupied by Allied forces and subject to US cultural influences (Robertson 2007). In 1951, a manga and anime character named 'Tetsuwan Atomu' (Mighty Atom), known as 'Astro Boy' in the West, became Japan's most famous robot. Astro Boy was created by Japanese artist Osamu Tezuka and was seen as a post-nuclear technology. The idea was that robots should help to bring out the best of our humanity in response to the nuclear age. The development of robots in Japan thus follows a parallel, but very different, course to that of Euro-American imaginaries. As anthropologist Christal Whelan (2014: 85) observes, 'robots in Japan were never conceived as enemies; they are friends and companions for life, tied intimately to humans in what anthropologists describe as fictive kinship'.

Computing and Cybernetics

In 1937, while studying for his PhD at Princeton University, British mathematician Alan Turing published the paper 'On Computable Numbers, with an Application to the Entscheidungsproblem' (Turing 1937). In it, he proposes a hypothetical device, later known as a 'Turing Machine', which could (in theory) solve any problem if it was described abstractly, as an algorithm. This suggested that any logical process done by a human could also be done by a machine. Ultimately, this established not only the enduring human–computer analogy but also led to the creation of the modern digital computer. During the Second World War, Turing led a group of British code breakers to develop a device to unscramble German codes and cipher systems, leading to the development of the first British electronic digital programmable computer, called 'Colossus', in 1943 (Russell and Norvig 1995). The following year, the ENIAC (Electronic Numerical Integrator and Computer) was completed at the University of Pennsylvania in the US.

Once again, these new technological breakthroughs brought the promise of developing a machine that could think like humans. In 1942, an invitation-only event called the 'Cerebral Inhibition Meeting' was held, focusing on the topic of hypnotism and conditioned reflex. These meetings were attended by anthropologists Gregory Bateson and Margaret Mead, as well as neurophysiologist Warren McCulloch, social scientist Lawrence Frank, psychoanalyst Lawrence Kubie and physiologist Arturo Rosenblueth (American Society for Cybernetics 2025). At the meeting, Rosenblueth presented on topics such as 'teleological mechanisms', 'circular causality' and 'feedback', which was subsequently published along with co-authors Norbert Wiener and Julian Bigelow in their article 'Behaviour, Purpose and Teleology' (Rosenblueth et al. 1943). The article explores the topic of purposeful machines and proposes that machines could be analysed using the approach of behaviourism from psychology, rather than the traditional functionalist approach of engineering.

Following on from this initial meeting, the first of what became known as the Macy conferences was held the next year. Its stated goal was to bring together leading thinkers to use the new technological advances to establish 'a general science of the workings of the human mind' (American Society for Cybernetics 2021). The conferences would run for the next eleven years, consisting of 160 conferences in total, and led to the establishment of the field of cybernetics. Attendees over the years also included mathematician and engineer Claude Shannon, logistician Walter Pitts, roboticist and neuroscientist William Grey Walter and mathematician, physicist and engineer John Von Neumann. Shannon's work had shown that switching circuits could solve the same problems as Boolean algebra, and McCulloch and Pitts's (1943) paper had proposed a computational model of a neuron. Together, they appeared to connect mechanism, Boolean algebra and animal physiology. In his book *Cybernetics or Control and Communication in the Animal and the Machine*, Wiener (1949) outlined his view of a world made up of complex feedback loops in which sensors and actuators exchange signals and information. Around the same time, Shannon (1948) established the field of 'Information Theory', defining information as a probability function with no dimension, no materiality and no necessary connection with meaning. The conviction that united all attendees was that computing mechanism and human thought were analogous (Hayles 1999).

The claims of the cyberneticians appeared to be supported by the parallel development of small, automated, analogue robotic devices, providing a tangible instantiation of the theories they were proposing. The most famous of these were Walter's 'tortoises'. Walter was a neurophysiologist and developed the robots as a means to try to understand the mechanics of the brain. The robots were named tortoises due both to their appearance, but also to evoke the mock turtle in *Alice in Wonderland*, who 'taught us', as they might reveal the secrets of life and of thought. These small robots, which he named Elmer (ELectroMEchanical Robot) and Elsie, had three wheels, blinking lights, and were light sensitive. When they encountered a mirror, they responded by displaying surprisingly lifelike behaviour, which Walter described as 'flickering, twittering, and jigging like a clumsy Narcissus' (Walter 1963: 115). Using behaviourist logic, he argued that this 'might be accepted as evidence of some degree of self-awareness' (1963: 128–9). Walter's intuition, which he believed the tortoises confirmed, was that simple mechanism could lead to complex behaviours through complex connections.

Wiener and Shannon also had robots. Wiener's robot was a small tricycle cart with two photocell sensors facing to the front. The output from the photocells was connected to the tiller controlling the front wheel. Depending on the output voltage, the cart either steered towards the light, like a 'moth', or away from it, like a 'bedbug'. Shannon's robot was described as an 'electronic rat' that exhibited goal-seeking behaviour and was able to find its way through a maze and appeared to 'learn' by trial and error. These physical instantiations of the cybernetic programme reiterated, and appeared to provide scientific and empirical evidence for, the imminent potentiality of formalizing human and animal thought and action. They also incorporated both human and physical sciences, integrating them under a unifying 'information science'.

Throughout the 1940s and 1950s, science fiction continued to parallel technological fields. Science fiction writer and biochemist Isaac Asimov wrote prolifically, including the *Foundation* series and *I, Robot*, a collection of short stories that focus on the moral predicaments of interactions between humans and robots. In one of the stories, the *Runaround*, he devised the 'Three Laws of Robotics' (Asimov 1942):

1. A robot may not injure a human being, or through inaction allow a human being to come to harm.
2. A robot must obey the orders given it by human beings except where such orders would conflict with the First Law.
3. A robot must protect its own existence as long as such protection does not conflict with the First or Second Law.

His stories focus on the unintended and counter-intuitive ways in which the robot applies the laws. As we will see, these fictional principles would come to have an outsized influence on the robotics community as they started to grapple with the issues of ethics and regulation for robotics in the next century.

The Birth of Artificial Intelligence

In 1955, a group of four scientists decided it was time to come together to settle the problem of 'intelligence' for good. They were John McCarthy, assistant professor of mathematics at Dartmouth College, Marvin Minsky, Harvard Junior Fellow in mathematics and neurology, Nathaniel Rochester, manager of information research at IBM and Claude Shannon. In their proposal for funding, they wrote:

> We propose that a 2-month, 10-man study of artificial intelligence be carried out during the summer of 1956 at Dartmouth College in Hanover, New Hampshire. The study is to proceed on the basis of the conjecture that every aspect of learning or any other feature of intelligence can in principle be so precisely described that a machine can be made to simulate it. An attempt will be made to find how to make machines use language, form abstractions and concepts, solve kinds of problems now reserved for humans, and improve themselves. We think that a significant advance can be made in one or more of these problems if a carefully selected group of scientists work on it together for a summer. (McCarthy et al. 1955: 12)

In a deliberate move to break from both cybernetics and behaviourism, McCarthy proposed this as a new field of 'Artificial Intelligence', or AI, simultaneously laying the foundations for a new psychological theory of 'cognitivism'. The Rockefeller Foundation provided funding for the event and other like-minded researchers were invited, including Trenchard More and Arthur Samuel of the IBM Corporation, and Oliver Selfridge and Ray Solomonoff of Massachusetts Institute of Technology (MIT). Once again, what bound this group together was a firm belief that the essence of what is human could both be replicated by a machine and understood in a formal and scientific way.

Although they did not succeed in their stated goal, they nonetheless established a research programme that would dominate both psychology and AI for the next thirty years. Despite this, the biggest technological achievement to come out of the conferences went almost unrecognized at the time (McCorduck 2004). This was the work of two minor participants, Herbert Simon and Alan Newell, who had met when they were both consulting at the military think-tank,

RAND Corporation. Simon, a professor of political science at Carnegie Tech, became interested in computers through Newell's work. Their first collaboration, along with computer programmer Cliff Shaw, was on a program that could process symbol structures using 'heuristics'. The resulting computer program, 'Logic Theorist', was presented at Dartmouth. Although their contribution is now seen as a seminal moment in the history of AI, it received a muted response from the attendants at Dartmouth at the time. Precisely why is the subject of debate. According to Marvin Minsky, the other participants had not connected their work so closely to human psychology and 'the analogy between human and artificial intelligence was not generally accepted' (quoted in McCorduck 2004: 157). Despite the muted reception, there was no such reticence from Newell and Simon about their own work. They were convinced that they had succeeded in creating a machine that could think like a human: '[We] invented a computer program capable of thinking non-numerically, and thereby solved the venerable mind-body problem, explaining how a system composed of matter can have the properties of mind' (Simon 1996: 190).

The authors claimed their achievement was as central to understanding the human mind in the twentieth century as Darwin's principle of evolution by means of natural selection was to understanding biology in the nineteenth century. They claimed that: 'there are now in the world machines that think, that learn and that create. Moreover, their ability to do these things is going to increase rapidly until – in a visible future – the range of problems they can handle will be coextensive with the range to which the human mind has been applied' (Simon and Newell 1958: 8). Logic Theorist would go on to prove thirty-eight of the first fifty-two theorems in Whitehead and Russell's *Principia Mathematica*, and even find a more elegant proof than one of the originals (Crevier 1993: 46). It has subsequently become known as 'the first AI programme'. It also introduced several concepts which would define AI research problems for the following decades, including reasoning as search, the use of heuristics to guide search and a new programming language called IPL, which would form the basis for McCarthy's AI-programming language, Lisp, in 1957.

Despite these breakthroughs in specific areas, Logic Theorist never managed to solve problems outside of the narrow problem area for which it had been designed. Nonetheless, it was Simon who, more than anyone else, pushed and ultimately legitimized the brain–computer analogy and thinking of humans as information processors (McCorduck 2004). The direct connection between machines and humans, which prior to the conference was still being shunned by serious scientists, was finally legitimized and AI was firmly linked to science.

In 1958, the US Defense Department's Advanced Research Projects Agency (ARPA) was formed in response to Russian technological advances, such as Sputnik. The agency was interested in AI research and other 'far-out projects as insurance against unwelcome technological surprises' (Moravec 1995: 21). They were particularly interested in two areas of AI: its relevance to human psychology and its physical instantiation in robots. The tight coupling of human and computational thought processes brought about by military interest and the AI programme at Dartmouth produced not only the field of AI (conceptualized at the time as symbolic information processing) but also the subsequently dominant paradigm of cognitive psychology. By the 1960s, the science of machines and the science of humans had effectively merged. The symbolic reasoning programme established at Dartmouth would come to dominate AI research. Alternative approaches were sidelined, including early 'connectionist' research, such as Pitts and McCulloch's work on neural networks.

The Intelligence Race

In the period after Dartmouth, funding and resources were concentrated in a small set of institutions and researchers in the US. Over the following decades, ARPA, now DARPA, poured millions of dollars into the four main research centres of Massachusetts Institute of Technology (MIT), Carnegie Mellon (then Carnegie Tech), Stanford University and Stanford Research Institute (SRI) with its stated aim to 'fund people, not projects', allowing them to spend the money in any way they wanted to (Crevier 1993). McCorduck (2004: 261) writes, 'though each project had its own flavor, the general aim was the same – to produce some sort of independent agent that would function in the real world, or at least a somewhat impoverished real world'.

At Carnegie Tech, Newell and Simon's next project was to try to develop a machine that could demonstrate a more 'general' human intelligence. Thus, in 1959, they followed up the Logic Theorist with 'General Problem Solver', or GPS. GPS aimed to develop a more rounded example of human intelligence, using techniques such as means-ends analysis, planning and selective trial-and-error, and this project remained a research focus until 1968 (Crevier 1993; McCorduck 2004). GPS proved successful on a number of specific tasks, including logic problems and puzzles, but none could be said to represent anything like generality. Instead, they claimed it had 'pretences to generality', thus admitting to its limitations, while not closing off the possibility for its eventual realization and thereby shutting down funding (McCorduck 2004: 249).

Researchers in the other centres were less convinced about the need for extensive knowledge of how the brain works in order to develop machine-based intelligence (Crevier 1993). At MIT, John McCarthy and Marvin Minsky founded the MIT Artificial Intelligence group, although they were not agreed on the best path forward. McCarthy believed that AI would be realized through the application of formal logic. Minsky believed in trying out anything that might work. McCarthy left MIT in 1963 to establish a group at Stanford just as MIT's AI group had received a large DARPA grant. At MIT, a new programme called 'Project MAC', or 'machine-aided cognition', was established, subsequently becoming the MIT AI Lab in 1970. The same year the SRI split from Stanford University, partly in response to student protests and objections to military funding by DARPA, and became 'SRI International'. Minsky's strong personality came to dominate the research programme at MIT, forming a devoted band of research students and an ad hoc approach to projects (Crevier 1993). His early projects concentrated on various forms of natural language processing and search (Brooks 1991a: 6).

Among the many projects initiated at MIT during these years was a conversational computer programme called 'ELIZA', created by computer scientist Joseph Weizenbaum. ELIZA was an 'artificial psychiatrist' and used pattern matching and substitution to create a superficial simulation of a natural conversation between humans. Sherry Turkle (2011), an MIT student at the time, recounts interacting with the program. As she reports, despite being aware of ELIZA's superficial constructions, after a few generic openers, students would invariably start to share their intimate details and concerns with it. While at the time this did not concern Turkle, Weizenbaum himself was appalled. Much of his subsequent work was dedicated to warning people about the dangers of conflating humans with machines, and his concerns about the degree to which we have made the world like a computer. He writes: 'Ultimately a line dividing human and machine intelligence must be drawn. If there is no such line, then advocates of computerised psychotherapy may be merely heralds of an age in which man has finally been recognised as nothing but a clock-work' (Weizenbaum 1976: 8).

In 1963, computer scientist Seymour Papert arrived at MIT, having met Minsky earlier at a symposium in London. Papert was interested in psychologist Jean Piaget's work, specifically developmental intelligence in children, and, on his arrival, Minsky also become interested in these areas (McCorduck 2004). Both of them had strong personalities; they courted media attention, worked with Hollywood on movie productions and became known for their witty aphorisms and wild predictions. Together, they initiated new research programmes that were focused on theories of computation, robotics, human perception and child psychology (Crevier 1993).

One of the areas that Minsky and his team worked on was computer vision, attempting to translate the image captured on a television camera into something the computer can describe. Initially, he hired an undergraduate student to solve it over the course of the summer. In the end, it would take another thirty years and the advent of digital cameras before this was achievable (Crevier 1993). Minsky and Papert also became interested in 'Blocks Micro Worlds', a simplified world of geometric forms, such as pyramids, squares and rectangular blocks, in which computers could be programmed to interpret images, manipulate blocks, answer basic questions and move about. Using the Blocks Micro World, they tackled the problem of hand–eye coordination, combining vision programmes with robotic arms. Crevier (1993: 92) describes it as follows:

> The arm had a moving shoulder, three elbows, and a wrist; it used fourteen hydraulic cylinders for muscles. Before attempting to grab any blocks, the robot would hold its hand in front of the camera and wave it a little to see whether it really was itself. The computer then adjusted the coordinate system used in the image to make it correspond with the coordinates of the hand.

By connecting sensors (cameras) to actuators (robot arms) and using computing mechanisms to control problem solving, they had built machines that could be said, at least to a certain degree, to 'see', 'plan' and 'act'.

Over at Stanford, McCarthy had started the Stanford Artificial Intelligence Lab, or SAIL. His goal was to build a fully intelligent machine within a decade (Moravec 1995: 20). The DARPA-funded project aimed to combine the latest capabilities in hardware with those in software, specifically computer vision and natural language processing, to see how close they could come to creating a robot that would emulate human behaviour and cognition. According to Charles Rosen, who led the project, the US Department of Defense agreed to fund it because they thought that the robot might be able to act as a mechanical spy (McCorduck 2004). The result was 'Shakey', the first real attempt to build a general-purpose 'cognitive' and humanoid robot, combining logical reasoning and some degree of autonomous movement. Shakey consisted of three parts; a wheeled cart at the bottom, a cabinet in the middle which held the computer processors, and, at the top, a TV camera, a range finder and a radio-linked antenna. Shakey could 'perceive' its world, 'plan' how to achieve a goal and 'act' in a physical world to carry out the plan. However, as Crevier (1993) points out, Shakey's physical world was deliberately constructed to be a simple as possible. It was a life-sized blocks world consisting of seven rooms connected by eight doors and containing square boxes. In this environment, Shakey could follow instructions given by a keyboard in simplified English and then negotiate its way around the room and perform an action, such as moving or stacking a box.

The project was not a success from a technical perspective. As Carnegie Mellon roboticist and 'futurist' Hans Moravec (1995: 15) later observed, Shakey was 'impressive as a concept but pitiable in action. Each move the robot made, each image captured by its camera, consumed about an hour of computer time and had a high likelihood of failure.' A person walking across Shakey's field of view could immobilize the robot, sometimes for days. The team named the robot Shakey because it shook and shuddered as it moved (DARPA 2025). As McCorduck (2004: 268–69) observes, 'Shakey showed that you could not, for example, take a graph-searching algorithm from a chess program and hand-printed-character-recognizing algorithm from a vision program and, having attached them together, expect the robot to understand the world.' By the time McCorduck saw Shakey, 'he was ... a sad sight, immobile in a corner'.

Despite its limitations, however, Shakey was hugely popular. The *New York Times* featured the robot in 1968, and *Life Magazine* and *National Geographic* in 1970, with *Life Magazine* dubbing it 'the first electronic person' (Kuipers et al. 2017: 97). Crevier (1993: 96) reports MIT researchers' consternation with some of the more elaborate claims made by the article, including a supposed quote by Minsky that 'in from three to eight years we will have a machine with the general intelligence of an average human being', which, he says, Minsky vehemently denied.

Outside of the US, an AI laboratory was set up in Edinburgh University in the UK, in 1965. Two experimental robots were built there between 1969 and 1976, 'Freddy' and 'Freddy II'. Freddy integrated vision, manipulation and intelligent systems. However, both robots were subject to the same problems as Shakey. As McCorduck (2004: 268) relates, Bernard Meltzer of Edinburgh University derisively called Freddy a 'feeble creature'. The only serious rival to US dominance in robotics was Japan. In the 1970s, the Japanese government committed $200 million towards the development of robots. The WABOT project at Waseda University aimed to build the world's first full-scale biped humanoid intelligent robot. WABOT-1 was released in 1973, and was able to play the organ, walk and communicate through an artificial mouth. It had sensors and hands to grip objects and move them.

By the end of the 1960s, major AI research programmes were producing AI programs that 'proved theorems in geometry, solved problems from intelligence tests, algebra books, and calculus exams, and played chess, all with the proficiency of an average college freshman' (Moravec 1995: 8). Despite the technical breakthroughs and successes in these specific areas, however, the major problems that had dogged the field from the beginning continued. 'General' intelligence continued to elude AI practitioners.

AI Winters

In 1965, philosopher Hubert Dreyfus was hired to spend the summer at RAND. His brother, Stuart, was a computer programmer and engineer there, and had just received his PhD. Based on his observations at RAND, Dreyfus (1965) wrote a scathing attack on the field of AI which would become his book *Alchemy and Artificial Intelligence*. In it, Dreyfus compared the new science of AI to alchemy and mocked what he considered the absurd predictions of AI researchers. Dreyfus's argument, based on a phenomenological understanding of perception and intelligence, was that humans rarely use logic for problem solving. It did not go down well. RAND considered repressing the article (Crevier 1993). Papert (1968) wrote a derisive refutation of Dreyfus and his work, calling it 'irresponsible' and 'wrong'. Simon suggested that Dreyfus was using his connection to RAND to give himself unwarranted credibility (McCorduck 2004). According to Dreyfus himself, no one at MIT, other than Weizenbaum, would dare be seen having lunch with him (Dreyfus and Dreyfus 2009: 8–9). Cynically, in 1967, Papert arranged a chess match between Dreyfus and an AI chess program called 'Mac Hack', which the AI program won (Crevier 1993). This incident was used to further discredit Dreyfus's argument within the community. Nonetheless, as Crevier (1993: 125) acknowledges, 'time has proven the accuracy and perceptiveness of some of Dreyfus's comments'. Both Crevier and McCorduck claim that while Dreyfus may well have been right, it was his 'tone' that estranged people.

Other criticisms at the time were more difficult for the AI community to elide. As we have seen, despite advances in specific domains, AI and robotics research in the 1960s was mostly a story of disappointment and unrealized ambitions. In 1973, the UK government commissioned a report on the state of AI in the UK (Lighthill 1973). Its author, James Lighthill, gave a devastating critique of the field. Any successes, he claimed, such as list-processing languages, were no better than the research done in regular computing laboratories. Lighthill dismissed the field of robotics as one that was influenced by human imagination and fiction as much as by science. He even suggested a 'pseudo-maternal' relationship between robots and their builders, in order to compensate for the lack of the female capability of giving birth to children. He went on to assert that 'in no part of the field have the discoveries made so far produced the major impact that was then promised' (Lighthill 1973: 8). Robotics, as a bridging activity between automation and 'intelligence' was disappointing, its proponents delusional:

> When able and respected scientists write in letters to the present author that AI, the major goal of computing science, represents another step in the general process of evolution; that possibilities in the nineteen-eighties include an all-purpose intelligence on a human-scale knowledge base; that awe-inspiring possibilities suggest themselves based on machine intelligence exceeding human intelligence by the year 2000; when such predictions are made in 1972 one may be wise to compare the predictions of the past against [today's] performance. (Lighthill 1973: 13)

Additionally, Lighthill (1973: 11) noted that benefits 'have flowed primarily to the science of psychology: in fact, a new range of attitudes to psychological problems has been generated'. The report essentially ended funding for AI in the UK and Europe and solidified the ongoing reductions and tightening up of restrictions by DARPA in the US. This period of time has since become known as the first 'AI winter', coined in 1984 at the annual meeting of the American Association of Artificial Intelligence.

Perhaps it is no coincidence, then, that envisioning human-like machines once again fell to fiction. In 1977, the first *Star Wars* movie was released, introducing the lovable robots C-3PO and R2-D2 to the world, setting revised expectations and inspiring a whole new generation of roboticists. Although presenting a wildly different view of robots in society, Maria, the anti-hero of *Metropolis*, provided the inspiration for the charming, biped C-3PO (Star Wars 2014).

In Japan, in 1982, contrary to events in Europe and the US, the Japanese Ministry of International Trade and Industry (MITI) launched 'the Fifth Generation' project, underscoring their previous commitment to developing humanoid robots (Nilsson 2010). In 1984, Waseda University released 'WABOT-2', followed in 1985 by WASUBOT, a keyboard-playing, musical humanoid which was exhibited at the opening ceremony of the Tsukuba International Science and Technology Exposition that year (Takanishi 2019). Also at the exposition, Hitachi Ltd released the WHL-11, a biped robot that could walk. The following year, Honda released their first of seven biped robots, 'Experimental Model 0'. Not wanting to be outdone, both the UK and the US renewed funding. DARPA decided that robot navigation was sufficiently advanced and the number of mobile robot projects increased dramatically as funding once again materialized (Moravec 1995: 21). In the US, DARPA founded the Strategic Computing Initiative, and in 1983, the UK invested £350 million in the Alvey project for massively parallel computer processing.

Throughout the 1970s and 1980s, it was predominantly the corporate Expert Systems market that had sustained the field (Russell and Norvig 1995). However, as desktop computers became more powerful and less expensive, these machines proved too expensive to maintain and, in 1987, the market collapsed overnight. Japan's fifth-generation project had also failed to meet expectations. Investors became disillusioned and funding for AI projects was once again withdrawn: the second AI winter.

The New Robots

The disappointments and limited successes of the symbolic information programming approach meant that now alternative and experimental approaches could be trialled. One of the problems, it was theorized, was the failure of AI systems to deal with 'common-sense' knowledge, a problem that John McCarthy had first articulated in 1959 with his 'Advice Taker' programme. In 1989, McCarthy (1989: 1) defined common-sense knowledge as 'the basic facts about events (including actions) and their effects, facts about knowledge and how it is obtained, facts about beliefs and desires. It also includes the basic facts about material objects and their properties.'

In the mid-1980s, a professor at Stanford, Douglas Lenat, decided to tackle the common-sense knowledge problem head on. He launched project 'Cyc' to capture all of the knowledge and rules about the world, including 'facts' or 'assertions' such as 'nothing can be in two places at once' and 'animals don't like pain' (Crevier 1993: 240). The project was expected to last for 'two person-centuries' and is still ongoing. Originally, as Nilsson (2010: 447) reports, Lenat had thought he would need 'a couple of million assertions'. By 2010, Lenat had revised the number to two hundred million. Dreyfus was, perhaps not surprisingly, highly sceptical of the project. His critique, rooted in phenomenology, argues that common sense is not based on 'context-free entities and their relationships', rather, on 'knowing-how rather than knowing-that' (Dreyfus 1992: xviii, xxvii).

Around the same time, another radically different approach to AI was initiated at MIT. Roboticist Rodney Brooks started to look at older, cybernetic models such as Walter's tortoises, which had showed that apparently purposeful and complex behaviour could emerge from simple, non-computational mechanisms. Brooks experimented with various types of robot cognition, including investigating the connection between sensorimotor skills and intelligence, and cutting out the computation altogether (Brooks 1990). Brooks set his robots loose in the lab, with sensors directly connected to actuators and little computation. He noticed, as Walter had previously, that the robots responded directly to the environment and were able to display apparent lifelike and purposeful behaviour. 'It soon became apparent', Brooks (1991a: 16) writes, 'that the dynamics of the interaction of the robot and its environment are primary determinants of the structure of its intelligence.'

Brooks (1990, 1991a, 1991b) is a harsh critic of the traditional symbolic approach, claiming that the trend in computer architectures and models over the previous thirty years had a strong and misguided influence on our models of thought. Brooks (1991b: 12–14) calls approaches that are solely concerned with abstraction 'a dangerous weapon', 'self-delusion' rather than good science'. Further, he argues that the traditional approaches relied, both implicitly and also explicitly, on what he called 'folk understandings' of human and animal behaviour. Instead, he argues that 'real biological systems are not rational agents that take inputs, compute logically, and produce outputs. They are a mess of many mechanisms working in various ways, out of which emerges the behaviour that we observe and rationalize.'

More recently, Brooks (2002: 36) has criticized the way that intelligence is defined in the field, observing that 'intelligence in the early days of AI was thought to be best characterized as the things that highly educated male scientists found challenging'. He points out the, now commonplace, observation in the field that seemingly simple activities are more difficult to automate than what are considered 'higher' order faculties, such as abstract, logical and mathematical reasoning. Such simple activities might be 'distinguishing between a coffee cup and a chair, or walking around on two legs, or making aesthetic judgements' (2002, 36). This has become known as 'Moravec's paradox'. In Moravec's (1995: 15) words: 'We are all prodigious Olympians in perceptual and motor areas, so good that we make the difficult look easy. Abstract thought, though, is a new trick, perhaps less than 100 thousand years old. We have not yet mastered it. It is not all that intrinsically difficult; it just seems so when we do it.'

Although it remains understated in his work, Brooks's views were influenced by phenomenology, in particular by Dreyfus's critique, the work of his students Philip Agre and David Chapman and, indirectly through them, the work of anthropologist, Lucy Suchman. Agre and Chapman (1987) had developed a computer program called 'Pengi' in which they show that planning, traditionally conceived as the mechanical execution of an explicit representational model of the world, can instead be designed without explicit models. According to Brooks (1991a: 2), Agre and Chapman's redefinition of 'routine activity in a relatively benign, but certainly dynamic world' was one of the core ideas that changed the course of his research. Instead of a focus on problem solving, planning or representations that 'rely on a semantic correspondence with symbols that the agent possesses', action was defined through interactions between the agent and the world (Brooks 1991a: 2). Thus, Brooks and his team were led to 'hypothesize (following Agre and Chapman) that much of even human level activity is similarly a reflection of the world through very simple mechanisms without detailed representations' (Brooks 1991b: 7). For Brooks, the core implications of these concepts were: first, a refutation of the representational approach, and second, the fact that 'intelligence' emerges from the interactions with the environment involving 'simple' mechanisms on the part of the agent.

As a result, Brooks (1990) developed a new robot architecture that he called 'subsumption architecture'. Instead of a traditional centralized, top-down system, the new system is decomposed into smaller pieces and distributed around the robot. It has a layered model of increasing complexity: for example, the 'bottom' layer might be a simple control system focused on avoiding obstacles, whereas a more complex, 'higher' level might consist of a control system with the capability to identify something significant in the environment. When not occupied by the more complex activity, the robot can simply move around using very little computation. By drastically simplifying the computation, reaction times also improved dramatically. This allowed them to build and test robots in uncontrolled, more 'natural' environments, without the problems that had been experienced by earlier robots like Shakey and Freddy. A number of new robots were built using the new architecture, including 'Allen' (after Newell), 'Simon' (after Herbert) and 'Ghengis' (after Khan). As Brooks (1990: 5–6) writes, 'Allen would happily sit in the middle of a room until approached, then scurry away, avoiding collisions as it went.' The more sophisticated Herbert could 'wander around office areas, go into people's offices and steal empty soda cans from their desks'. The six-legged Ghengis could walk over rough terrain. Building on these successes, Brooks hoped that he would build human-like intelligence incrementally, much like evolution.

In a new project, Brooks decided to pursue what he called the 'Holy Grail' of the AI community: an android, or 'autonomous robot with humanoid form and human-like abilities' (Brooks et al. 1999: 1). In 1993, Brooks put together a team that included philosopher and cognitive scientist Daniel Dennett to work on an upper-torso humanoid robot called 'Cog'. Cog's system was to be based on four alternative 'essences of intelligence', gleaned from work in developmental psychology, ethology, systems theory, philosophy and linguistics that the team claimed had been 'discarded' by traditional symbolic approaches. These were: physical embodiment, integration of multiple sensory and motor systems, a developmental structure allowing for incremental learning, and social interaction. Cog's perceptual system included traditional visual and auditory sensors (cameras and microphone), but also a 'vestibular, tactile and kinaesthetic system' developed to allow for advanced physical orientation within the world and in relation to other objects (Brooks et al. 1999: 9, 1). The task the team had set themselves was to find a skill decomposition that maintains the complexity and richness of the behaviours represented while remaining simple to implement and construct. This, they believed, would allow them to contribute both to engineering and to science, that is, to robotics and to 'the scientific goal of understanding human cognition'. As Adam (1998: 147–78) observes, the team was strongly influenced by evolutionary biology: 'Cog was programmed to recognise its "mother's" face (a post-graduate student), and will be designed to learn, so that future descendants of Cog can retrace the steps of millions of years of evolution in a few years of laboratory experiments.' Despite Brooks's earlier insights related to a 'benign world', according to Adam, the evolutionary style of robotics incorporates '[a] view of nature red in tooth and claw', in which '"[s]uccessful" robot control systems can contribute genetic material to become "parents"' while unsuccessful ones 'are "discarded"' (Adam 1998: 154).

Ultimately, Cog's key scientific achievement was to highlight its own shortcomings. The Cog project did not succeed in its aim to advance our understanding of human intelligence or verify the four 'essences' of intelligence. With regard to 'physical embodiment', the project's website admits, '[s]ince we can only build a very crude approximation to a human body there is a danger that the essential aspects of the human body will be totally missed' (CSAIL 2025). Agre (1997: 242) had made a similar observation in reflecting on the Pengi programme: '[i]t is hard to say which aspects of human embodiment and acculturation are necessary, either by definition or as a practical matter, for the human forms of intentionality'. As described on MIT's Computer Science and Artificial Intelligence Laboratory (CSAIL) web page for Cog, the key learning from physically embodied robots is the dramatic reaction they provoke in the people who interact with them. This, they reason, might have the advantage of encouraging natural human interaction from which the robot could 'learn'.

Building on this insight, Cynthia Breazeal, one of Brooks's students, led a group at MIT to create Cog's successor in the late 1990s: a social robot called 'Kismet'. Kismet had a narrower remit than Cog and was specifically designed to produce an affective response in people (Breazeal 2002). Kismet has a large head and no body, with exaggerated features to simulate emotion, including two large eyes, eyelids and bushy eyebrows, as well as lips and a jaw. Its head has pointy ears made out of paper, which make it look like a gremlin. At the time, it was equipped with audio, visual and proprioceptive sensors to allow it to recognize and respond to emotional cues.

Breazeal (2002: xii) draws on the idea of 'social intelligence' to underpin Kismet's scientific credentials, arguing that social factors play a crucial role in our evolution as a species and in our cognitive development. Breazeal also draws on human developmental psychology, in particular studies based on early infant–caregiver interactions. Kismet is thus conceived as an infant with infant-level social abilities. The childlike design serves the twin functions of drawing people to it, while also excusing any shortcomings in its interactive capabilities. As Suchman

(2007: 237) observes, 'the figure of the child in Euro-American imaginaries carries with it a developmental trajectory, a becoming made up of inevitable stages and unfulfilled potentialities'. Thus, this strategy 'simultaneously authorizes the continuation of the project and accounts for its incompleteness'. The idea was that by engaging people over time, Kismet's 'intelligence' system could be trained, and thus incrementally develop increasingly sophisticated interactions. Kismet's 'intelligence' is thus not coded in from the start but conceived as a blank slate and something that will emerge from these interactions.

Breazeal (2002: 1, 7–11) breaks down social intelligence into five discrete components, distributed between the robot and the observer, allowing it to 'understand us and itself in social terms'. First, following Brooks, a socially intelligent being is 'embodied and situated'. Second, they should be 'life-like' and 'believable'. Third, they must be 'understood'. This means they must be perceived as social, by displaying 'a capacity to give attention, emotion, expression and playfulness', as well as enabling a human to read their actions and intentions. Fourth, they should be 'human-aware', being able to identify a person, what they are doing and how they are doing it' and, finally, they should be imbued with 'simple learning strategies', like 'observation and feedback'.

Ultimately, however, neither Cog nor Kismet was able to display any demonstrable intelligent behaviour. By 2003, both robots were retired to MIT's science museum (Suchman 2019: 43). Yet Cog and Kismet's ability to evoke reactions in those who came into contact with them, as well as continuous media coverage, ensured that their influence far outweighed their capabilities. As well as those done by Suchman, Kismet was also the focus of a number of other prominent studies, including by Sarah Kember (2003) and Turkle (2011).

The lack of progress in AGI moved the focus of robotics from the inner workings of the robot as a replica of the human, to the robot as perceived from the outside, and as a human assistant. By the 2000s, funders developed an interest in the topic of human interaction and collaboration with robots. In 2001, the US National Science Foundation and DARPA sponsored a workshop on the new field of human–robot interaction (HRI), to help identify the issues and challenges in order to design 'synergistic teams of humans and robots where team members perform tasks according to their abilities' (Burke et al. 2004: 104). The new HRI field was thus established quite explicitly to engineer both the robot and the human. The steering committee featured Breazeal from MIT, professor of communication at Stanford University Clifford Nass, and roboticist Ronald Arkin from Georgia Tech.

Throughout the 1990s, Waseda University in Japan also released new robots focused on human–robot communication and interaction. Between 1993 and 2000, Honda experimented with a number of increasingly sophisticated and sleek-looking robots, culminating in 2000 with the release of 'ASIMO', named after Isaac Asimov, a small biped robot clad in white plastic, with a helmet and black visor, resembling an astronaut.

Other Japanese roboticists concentrated on making robots look as lifelike as possible. At Osaka University, Hiroshi Ishiguro created the lifelike 'Actroid' robot in 2003, with a female appearance. In 2006, he released the 'Geminoid' robot, modelled on himself. The robots can blink and appear to breathe and 'fidget'. Similarly, Hong Kong-based Hanson Robotics created the controversial 'Sophia' robot, who has since been granted 'citizenship' of Saudi Arabia and is the first non-human United Nations Development Programme's 'Innovation Champion' (UNDP 2017). Despite some limited automated communication capabilities, public appearances for these robots are teleoperated.

In 2005, MIT spin-off Boston Dynamics started to make robots with funding from DARPA. An early project was a four-legged 'canine-inspired' robot named 'BigDog', designed for military purposes. BigDog displayed extraordinary stability, strength and agility, but was considered too loud for use in combat (Degeler 2015). BigDog was followed up by the smaller 'Spot' and the humanoid 'Atlas'. Videos show these robots performing extraordinarily lifelike feats, such as opening doors, dancing, navigating down hallways and walking up and down stairs. To an observer, these displays give a striking impression of animacy and sentience that is not matched by their inner computational abilities. Nonetheless, these demonstrations did much to further the idea of imminent robot sentience.

In 2006, the new Japanese Prime Minister Abe announced plans for a 'beautiful (utsukushii) innovative, new, roboticized Japanese society'. As Japan anthropologist Jennifer Robertson (2007: 385) reports, the proposal, called 'Innovation 25', lays out plans for 'the roles that biotechnology and robotics will jointly play in securing the stability of both the Japanese economy and Japanese social institutions' by 2025. Robertson calls the vision that was laid out 'reactionary postmodernist', in which robots are designed to free up women's time so that they can devote more of it to having children and to compensate for the declining and ageing population, thus avoiding the need for immigration. Further, according to Robertson, these new Japanese robots herald an era of increased surveillance and control. In the US, Boston Dynamic robots are also coming under scrutiny. Spot was trialled by four different police precincts. In 2021, they were removed from the New York Police Department after a public outcry, including that from US representative Alexandria Ocasio-Cortez (2021), who criticized their use 'for testing on low-income communities of color with under-resourced schools'.

Despite the difficulties in developing AGI, as well as the ethical and regulatory challenges, robots as technologies are becoming increasingly prevalent. Robots are increasingly being used in manufacturing, in transport, in space exploration, farming, retail, care and service roles, bomb disposal, surgery and as weapons in war. In keeping with the evolutionary narrative that has become a part of the field, a number of my informants referred to the current proliferation of robots as a 'Cambrian explosion'.

Anthropology Theory and the Cyborg

Beyond warnings about the risks and limitations, anthropological theory has surprisingly little to offer to roboticists concerned with trying to build people. Despite its nominal focus on 'anthropos', anthropology is in fact leading the charge to eradicate its object. In order to understand why this is the case, it is helpful to look back to the 1970s and 1980s. At this time, a number of ethnographers turned their attention to studying scientific laboratories in the 'West', a radical departure from the traditional focus on 'primitive' cultures. This is what Bruno Latour (1993) calls a 'symmetrical anthropology', one in which both 'non-modern' and 'modern' cultures were subject to the same investigative methods. These laboratory studies revealed that, contrary to common assumptions, activities in the lab that were assumed to be linear and stepwise, instead involved 'tinkering' (Knorr-Cetina 2014) and 'bricolage' (Latour and Woolgar 1986; Levi-Strauss 1966). Scientific reasoning and decision making were shown to be 'indexical', context dependent and post hoc, rather than rational and predictable (Knorr-Cetina 2014). These insights even led some to conclude that nothing 'scientific' happens in scientific laboratories (Knorr-Cetina 2014; Latour 1983).

Not surprisingly, these conclusions met with pushback from the scientific community. In response, Latour and other scholars distanced themselves from their early overtly social emphasis and developed an alternative to either social or technological determinism. This was Actor Network Theory, or ANT, a novel techno-social arrangement, which emphasized not the entities in the network but the relations between them (Callon and Latour 1992). In a network, at least in principle, neither the social nor the technical elements are privileged. Instead of action and agency being a property of individual humans or objects, they emerge from the interactions between them. Gradually, symmetrical anthropology gave way to a wider principle of symmetry, a 'flat' or 'symmetrical ontology', in which it is not just modern and pre-modern humans subject to the same investigative methods, but also humans and 'non-humans'. Ultimately, Latour (2002, 2004) extends his symmetrical approach to encapsulate a view of non-humans as having a moral dimension and to bring them into the political sphere.

The principle of symmetry had much to offer anthropological theory. It was seen as a solution to the problematic Cartesian dualism, which separates mind from body and culture from nature. It was also seen a radical departure from humanism, essentialism and anthropocentrism, and thus offered an alternative to traditional Enlightenment rationality. The focus on relations, rather than entities, offered an effective counter to the predominant focus on objectivity, atomism and individualism. It was thus also enthusiastically adopted by scholars who took a more critical approach, reflecting on issues such as class, gender and race. For these theorists, observations of research in biology (Haraway 1989) and technology (Adam 1998; Kember 2003; Suchman 2007) revealed how boundaries are porous between nature and culture, humans and non-humans, and are constructed, rather than given. This led to the development of 'posthumanism', a response to the anthropocentrism and foundational assumptions of humanism (Haraway 1991; Hayles 1999).

Posthuman feminist approaches argued for a renewed consideration of 'materials' and things, echoing the Latourian calls for extending agencies beyond the human. The posthuman material turn also brought renewed emphasis on the body, particularly the female body, which, it was argued, had been omitted from traditional scientific accounts (Adam 1998; Alaimo 2010; Grosz 1994). Posthuman, relational and anti-essentialist perspectives, with their 'symmetrical' or 'flat' ontology, have remained a persistent theme in anthropological theory characterized by problematizing previously taken-for-granted distinctions between people and things, and a proliferation of non-essentialist constructs, such as networks, hybrids and cyborgs.

Donna Haraway's (1987) figure of the 'cyborg' is perhaps the most iconic formulation. The idea of the 'cyborg' originated with Clynes and Kline (1960), who had imagined a technologically enhanced human being who could survive in space. Haraway reimagines the cyborg as a means of disassembling traditional categories. Like Latourian 'networks', the 'cyborg' concept rejects rigid boundaries, focusing instead on relations and the possibility of reconfiguration. But much more than the network, Haraway's cyborg is primarily a political construct: an imaginative resource and new mythology representing a break with traditional narratives and origin stories. While Haraway recognizes that the original cyborg is a product of militarism, patriarchal capitalism and state socialism, she proposes that understanding technologies solely from the perspective of domination neglects the possibility for emancipatory politics that comes with the radical disruption and power that new technologies can bring. Instead, she proposes that the cyborg can be used to resist the 'the logic of appropriation, incorporation' and instead accommodate 'partiality, irony, intimacy, and perversity' (Haraway 1987: 9).

The cyborg is thus a way to articulate more positive configurations between humans and technology, as opposed to traditional critiques that insist on the threat and inevitable domination of society by technology. The cyborg is a vision for a future that can be forged from within the messy reality in which we now live, ensuring that critiques of technology do not adopt 'an anti-science metaphysics, a demonology of technology' and that allies for this collaborative reimagining might be found across all the sciences (Haraway 1987: 39). The cyborg is part machine and part organism, a creature of both reality and fiction, which does not emerge from nature, from women's bodies or from a pure, prelapsarian past. Haraway proposes that, through the figure of the cyborg and its troubling of traditional binaries (such as human and animal, organism and machine, man and woman), we might be able to disrupt the traditional concept of the human and learn 'how not to be Man, the embodiment of Western logos' (Haraway 1987: 31). Ultimately, she hopes to reorient feminist accounts to the positive potential of science and technology in order to blur and transform (oppressive) boundaries and identities and instead offer a utopian vision of a post-gender, post-essentialist world.

The expectation of imminent machine animacy is a foundational assumption of relational and posthuman scholarship, and a key argument for the dissolution of the human–machine boundary. As Haraway (1987: 11) writes: 'Late twentieth-century machines have made thoroughly ambiguous the difference between natural and artificial, mind and body, self-developing and externally designed, and many other distinctions that used to apply to organisms and machines. Our machines are disturbingly lively, and we ourselves frighteningly inert.' As well as the basis for arguing for the dissolution of boundaries, this assumption is also used to justify the increasing obsolescence of the human as a core category of ontological significance.

Boundary Resistance

By foregrounding technology, nature and embodiment, relational and posthuman approaches interrogate and problematize traditional boundaries between society and nature, humans and machines and mind and body. Both Latour (2004, 2014, 2015, 2017a, 2017b) and Haraway (2016) further develop their relational and posthuman approaches in the context of the climate crisis. These approaches challenge what are seen as the inherited anthropocentrism and radical individualism of traditional social and economic theory. Furthermore, they promise to bridge the ontological 'separatedness' between humans and nature. It is clear that the work of both of these thinkers is crucially important in thinking through these issues, and their work is central to much contemporary scholarship focused on this subject,

However, for researchers concerned with the climate crisis, the non-human or 'more-than-human' side of the dialectic is generally considered to refer to natural entities such as water, wind, earthquakes, animals and trees. By contrast, for ANT and the Latourian symmetrical view in particular, 'non-human' generally refers to technical objects and consumer goods. Indeed, through ANT, while human exceptionalism, intentionality and agency are thrown into doubt, objects are elevated to 'integral and active threads of an intersecting scientific, social and discursive world' (Conty 2018: 5), with the discourse of the human subaltern redeployed in defence of the non-human object (Fowles 2016: 12). There is traditionally very little interest in non-human animals in ANT and relational ontologies. And yet, it is the human–animal boundary that is most unstable and open for reconfiguration. Increasingly, studies from across scientific and philosophical fields are revealing animal capacities for traditional 'human' qualities, such as empathy, affect, care, sentience, vulnerability and self-affection (Browning 2023; Huopalainen and Jammaers 2024; Mellor 2019; Nussbaum 2024; Pérez-Manrique and Gomila 2018). These studies reveal animals engaged in nuanced empathic behaviour, such as showing sympathetic concern, experiencing personal distress in response to others' suffering and targeted helping even of unfamiliar others (Pérez-Manrique and Gomila 2018). In the symmetrical view, the unitary category of the 'non-human' belies a vast spectrum of heterogeneous entities and beings that have little in common, and erases differences between them.

Far from being neutral, then, dissolving boundaries between humans and technology has far-reaching implications for both. Equating technologies on an ontological, political and moral level with living organisms anthropomorphizes technologies, while objectifying living beings, including humans. As well as the objectification of humans and animals, the subjectification of objects leads to ignoring their material facticity: their physical form and durability, their historical and temporal specificity and their 'complex sensuality' (Henare 2003: 57).

Despite claiming to dissolve dualistic boundaries, underlying these relational and posthuman works is often an implicit mind–body dualism. This persists in a view of mind as disembodied and cognitive, in which human perception and understanding is the result of symbolic representations. In her book *How We Became Posthuman: Virtual Bodies in Cybernetics, Literature and Informatics*, Kathleen Hayles (1999: 3) acknowledges how, in the posthuman view, the human is configured 'so that it can be seamlessly articulated with intelligent machines'. This also has implications for attempts to focus on embodiment and the body in posthumanist scholarship. According to Hayles, the cybernetic construction of the posthuman has had the most impact in downplaying and erasing the role of the body and embodiment as an essential element of the human. This is, she maintains, a feature that cybernetic posthumanism has in common with the liberal humanist subject, whose universal human essence transcends the body, the body instead being its possession. In her feminist critique of AI, Alison Adam expresses similar concerns, worrying that 'in order to eschew essentialism we may end up disembodying the body' (Adam 1998: 135).

Finally, but importantly, within anthropological theory, there is a persistent fear of being charged with dystopian and Luddite views of technology. In order to take technologies and objects seriously, it has become necessary to eliminate any distinction between them and living things, or else be accused of anthropocentrism, essentialism or excessive anti-materialism. This can also lead to an unfounded optimism about the possibilities for transformation and revolution through technology. It is important to note, as Suchman (2007: 269–70) does, that symmetrical ontology arose in the social sciences and humanities, which had previously excluded 'facts of nature' and 'technology' from its analysis. Outside of these fields, however, 'the situation is in important respects reversed'. Indeed: 'Far from being excluded, "the technical" in regimes of research and development are centered, whereas "the social" is separated out and relegated to the margins. It is the privileged machine in this context that creates its marginalized human others.' Thus, within the social sciences and the humanities, the symmetrical view provides a productive and problematizing critique. Outside of these fields, by contrast, it serves both to extend the privileging of the technical over the social and to obscure important questions about the category of the human, as well as other animals.

In this chapter, I have shown that, in contrast to the view of robots as a cutting edge and futuristic technology, the robot's lineage stretches much further back into history; to ancient attempts to represent and reproduce life and human likeness. Indeed, robots are at the centre of centuries of attempts by mathematicians, philosophers and engineers to distil the essence of life and humanity, attempts that have consistently eluded them. Despite this, parallel deceptions, dissimulations, and fantastical portrayals in media and fiction have continued to exaggerate, romanticize and promote discourses of machine animacy. These misconceptions have been compounded by the intense visceral reactions that people experience when encountering the robot. Furthermore, the image of the human and the machine have been co-constructed, with each serving as a model for the other, and in recent times, the human–machine analogy continued to be carefully constructed and institutionalized. Anthropological theory, in particular the posthuman symmetrical approach, builds on this analogy to argue for the dissolution of the category of the human. In the chapters that follow, I start to trouble the human–machine analogy, echoing earlier calls from roboticists such as Weizenbaum, who understood what is at stake with erasing this boundary. However, rather than reconfirming human exceptionalism, I argue for a greater continuity between human and animals. In the next chapter, I start by exploring how these histories are materialized through the research sites, laboratories and wider institutional networks within which robotics work takes place, as well as how they are embodied in the robot. This will further uncover the gap between the reality and the fantasy, revealing a disconnect between the capabilities of the technology and the discourses of imminent AGI. I will also reveal the reality of the risks of robots and AI, which are simultaneously more mundane and more urgent.

Note

1. Everyone who interacts with Stevie switches between 'he' and 'it' depending on the context. In this book, I refer to Stevie with the pronoun that I attributed to him based on situation of the interaction. I also reflect on this in subsequent chapters.

CHAPTER 2

Making Robots, Making Worlds

Inside the Robotics Lab

Artificial Intelligence and Simulated Behaviour, Sheffield

My first formal introduction to the world of robots and robotics research was at the Artificial Intelligence and Simulated Behaviour (AISB) conference at Sheffield University in the UK in April 2016. My knowledge of the field at this point had been mostly theoretical, and I had assumed that the other participants would be traditional engineering types, working on discrete aspects of robotics technologies and chasing very narrow versions of 'intelligence'. Instead, I found a field exploding with creativity and excitement. Humanoid robots were not a far-off dream, but already operational. Huge advances in mechatronics and swarm automation were creating moving machines that looked and felt sentient.

The convention consisted of a wide variety of topics related to AI and cognitive science: from AI and games to embodied cognition and acting, HRI, depression, artificial sexuality and the social aspects of cognition. Thus, as well as engineers and computer scientists, I met philosophers, psychologists, theatre people, legal scholars, designers, artists, sex therapists and priests. It was clear that something was happening, and if not the creation of human-like intelligence, then nonetheless a gathering of creative, conscientious and energized people, organized around the image of the robot. My presence as an ethnographer, far from being considered out of place, was completely unremarkable. There was an openness and inclusiveness to the field that I would continue to experience throughout my fieldwork.

Although there were many different disciplinary tracks, I was one of the few people moving between them. The topics in each track were similar, as researchers grappled with the philosophical and social implications of contemporary robots, yet each track dealt with these questions in completely distinct ways. Professor of cognitive science, Margaret Boden, was the star of the 'Computational Creativity' track, in which she proposed Google's Deep Dreaming algorithm as the 'new collage', challenging any essential difference between human-created art and computer-generated models. In the 'Embodied Cognition, Acting and Performance' track, theatre scholars and philosophers reflected on the embodied nature of performance, suffering and empathy with robots. In 'Social Aspects of Cognition and Computing', speakers discussed the social, ethical and philosophical aspects of robots. The HRI track was by far the most 'techie'. Research groups, including a large contingent from MIT, presented their controlled experiments between robots and 'human subjects', measuring discrete social variables such as 'joint attention' and 'gaze following', 'turn taking', imitation, socially tuned perception, social guided visual search and human–robot 'personality matching'. The human 'subjects' whom they were targeting with their work on social robots were primarily children, people with autism and older adults.

Despite the variety of disciplines and tracks present at the AISB conference in 2016, most tracks had an 'ethics' focus of some kind. What was meant by ethics, however, varied widely. The HRI group had a guest speaker who spoke about the risks of empathy and attachment with robots. The more philosophically oriented tracks debated how well traditional philosophical concepts stood up to the problem of potential robot consciousness, including, for example, how seriously we should consider robot suffering or robot rights from the perspective of traditional analytical categories, as well as how we might recognize consciousness, or extend moral concern for such a hypothetical sentient machine. Notably, none of these debates acknowledged the lack of moral concern in society for the many other sentient beings we share the planet with. Sentient beings that display the autonomy, consciousness and feelings that participants considered the yardstick for attributing moral standing to robots. There was a lot of discussion about programming 'human values' into machines, including prolonged debates about the 'trolley problem',[1] with an uncritical assumption that there was, or could be, a universal and agreed-upon moral standard for all humans.

Another key concern was how to control a machine that was 'smarter' than us. Philosopher Nick Bostrom's (2003) 'paperclip maximizer', a thought experiment examining the 'control problem', formed the basis of many discussions. In this experiment, a super-intelligent machine is given the goal of producing paperclips. It is also given the ability to learn, so that it can develop ways to achieve its goal better. The AI is both single minded and ingenious, so it starts to appropriate resources from all other activities. Eventually, the AI turns everything in the world into paperclips, including using the carbon atoms from our human bodies. All of the ethical debates were characterized by disagreement, confusion and contradiction, as participants from different academic traditions, and none, struggled to understand each other.

That year, AISB was also the setting of a symposium to develop and debate the 'Principles of Robotics', or an ethics code for robots. It was also due to the symposium that I had the good fortune to find myself drinking *sake* with some of these senior roboticists, who would ultimately become some of my primary interlocutors in the UK and US. The principles had originally been created in 2010 at a funded retreat for a group of the UK's senior roboticists to promote acceptance of robotics by the general public, so that it would not face the kind of opposition that genetically modified organisms had faced. The group had decided to take it upon themselves to create design principles for 'regulating robots in the real world' (Boden et al. 2017). The original principles, developed in 2010, consisted of five rules and seven 'high-level messages', deliberately building on Asimov's three laws. The new principles were as follows:

1. Robots are multi-use tools. Robots should not be designed solely or primarily to kill or harm humans, except in the interests of national security.
2. Humans, not robots, are responsible agents. Robots should be designed and operated as far as is practicable to comply with existing laws and fundamental rights and freedoms, including privacy
3. Robots are products. They should be designed using processes that assure their safety and security.
4. Robots are manufactured artefacts. They should not be designed in a deceptive way to exploit vulnerable users; instead their machine nature should be transparent.
5. The person with legal responsibility for a robot should be attributed.

When drafting the original, the group had immediately come up against Asimov's first law: A robot may not injure a human being or, through inaction, allow a human being to come to harm. The problem being that the key funder of AI and robotics research is the US military. The first and most basic of Asimov's laws, then, had to be heavily qualified. In 2016, the original attendees were back to develop and debate the principles.

When I first read the principles, I was surprised by their vociferous pragmatism, a trait that also struck me about many of the participants with whom I became acquainted over the course of the conference.

At the symposium, there was a pervasive sense that the participants gathered were responsible for guiding the rest of society in their use of technology and thus to make the world safe. When they talked about the public, they considered questions such as 'should *we* allow people to use robots?' These people saw themselves as world builders and world makers, and, despite an acknowledgement of the need for greater input from other disciplines, had limited awareness of alternative knowledges outside of their inherited analytical frames. The conference was a fascinating and exciting entry to the field, and I returned home determined to chase up humanoid robots that, while animating all discussions at the conference, I still had not encountered.

International Conference of Social Robotics, Tsukuba

The following year, one of my papers was accepted at the International Conference of Social Robotics (ICSR) in Tsukuba, Japan. Many roboticists consider Japan the 'spiritual' home of human-like robots, and throughout my fieldwork, researchers spoke enviously of the acceptance of robots in Japanese society, their culture and Shinto beliefs, which made it acceptable to attribute spirituality to things, and the government support for the robotics programme. The most successful commercial robot, the therapeutic furry seal robot 'PARO', is of Japanese origin, although it is very far from what many people, at least in the Euro-American imaginary, would associate with the idea of a robot. The sleek and minimalist design of Honda's humanoid robot ASIMO provides the template for today's social and care robots. Despite this, Japan in fact trails behind South Korea and Singapore in terms of robot density and behind China in terms of operational stock (Kodate et al. 2022).

Tsukuba is about 50 km north of Tokyo, and one of the largest science and technology sites in Japan. It is also the location of the National Institute of Advanced Industrial Science and Technology (AIST), where the robot PARO was developed. Tsukuba is designated a 'Mobility Robot Experimental Zone', which allows for robotic experiments on public roads, and is marked by signs indicating as much across town. However, as also noted by Bender (2025: 52), this is not reflected by their actual presence on the ground, which was somewhat disappointing. Despite its promising cultural setting, the HRI and social robotics conference proper was focused narrowly on laboratory experiments of micro-interactions between robots and 'human subjects'. Papers included topics such as the impact of robot errors on people's trust, the influence of robot body shape on perceptions of gender and the influence of the robot's interaction style on user performance. Most studies used physiological measures, such as pupil diameter or heart rate, or statistical measures, such as Likert scales and other measures of 'user acceptance', methods that are limited to measuring only existing norms and majority expectations. The research thus tended to focus on finding statistically significant differences in the data, with an emphasis on differences between genders. There was a marked lack of critical engagement with the topics, or with the potential impact of reproducing social norms in the technology. The studies relied heavily on Wizard of Oz methods in lab environments, in which a researcher teleoperates the robot to emulate 'intelligent' behaviour, often covertly. In surveys, participants were often recruited via Amazon's Mechanical Turk, a crowdsourcing website for recruiting 'crowdworkers', which is discussed in more detail in Chapter 4. The effect of these approaches is to make a research field that is, as one of the pioneers of the field admitted to me privately, 'prematurely conservative'.

Due to this narrow focus, many HRI and social robotics conferences reserve alternative tracks and workshops for divergent themes that are inevitably also a part of the discipline. Thus, as well as micro-validations going on in the main conference, workshops included the kind of topics that threatened from the fringes to destabilize the serious and scientific image of the field, such as a delightful 'robots and religion' workshop. This workshop featured a transhumanist pastor who spoke about the hope of immortality through AI. There were presentations on the 'Tao' of robots, and a demonstration of the 'first Catholic robot', the 'Sanctified Theomorphic Operator', or SanTO. Japanese robot designer Tatsuya Matsui, of Flower robotics, spoke at length about the robot aesthetic, which, like a flower, can be delicate and beautiful, and in need of nurturing. At the conference, the vegan and vegetarian table during mealtimes was a particularly lively site of discussion and interaction. Throughout my fieldwork, it was notable the number of roboticists who followed a vegan or vegetarian diet, often for espoused ethical reasons. It was at this conference that I met two other key informants and subsequent collaborators, Daniel and Gerry, after which I became truly engaged in the field.

Trinity College Dublin

To get to the Robotics Lab at Trinity College in Dublin, you have to climb up a number of flights of stairs of a historical building, walk down a long corridor and through a new extension at the back. When I first went there, I marvelled at the historic location and aesthetic but the team were not feeling sentimental. The building did not represent innovative engineering or the futuristic feel that they felt they were part of creating. The small room that houses the lab is one of the few robotics labs in Ireland. Daniel, who leads the team, is an assistant professor and mechanical engineer, and the small team is predominantly made up of engineering post-docs, as well as a number of post-docs from computer and data science. For Daniel, the project is the culmination of a life-long dream of inventing something new and taking his place among the great inventors.

The lab is a large, L-shaped room with large, steel-mounted shelves all around the walls. The shelves are piled high with wires and mechanical parts, nuts and bolts, bits of plastic and wood, and tools and machines. On the right-hand side, as you come in, there are a number of high tables with machines on them, as well as a 3D printer. In the main part of the room, there are four long desks set against the wall with people working at computers, as well as a few large desks set at right angles to the wall, with high stools, and flat screens at the end and a camera for video conferences and meetings. In the centre, there are two sofas facing each other and a coffee station. The room is part workshop, part computer lab, with a dash of design aesthetic. The work of the lab is almost entirely focused on building a robot called Stevie.

Stevie is a full-bodied humanoid robot. The upper part of his body has a 'human-inspired torso' with two short arms ending in balls, and a head. The head, or 'social interface', is connected to the torso via a 'three-degrees-of-freedom' neck mechanism. This means that there are just three defined directions in which the neck can move using motors, including pitch (side-to-side) and roll (front to back). The front of the head, or face, has two separate digital displays to allow Stevie to display eight different emotions. Stevie's arms have two degrees of freedom at each shoulder. Stevie is 1.4 metres tall, about the size of a ten-year-old child.

Stevie's height and appearance are in keeping with the current accepted thinking on social robots. Research from Japan, particularly on Honda's ASIMO robot, suggests that the optimum height for a social robot is 1.2 metres (Hirose and Ogawa 2007). For Stevie, the researchers, 'through trial and error', a combination of needing to fit the parts and aiming for a reasonable stature, came up with 1.4 metres. As they explain to me, this is tall enough to 'have presence', but small enough not to be too imposing or intimidating or dangerous if it falls over. It also has the advantage of being at eye level with someone who is sitting down or in a wheelchair. Trying to get a robot, especially a wheeled robot, to bend over would add a huge amount of complexity, they explain. There is a large red button on the shoulder, which is a stop button, or 'kill switch': a safety feature. Stevie needs power to function, and lots of it. There are two very large batteries between the wheels at the bottom, which have been specifically designed to fit.

The fact that Stevie has wheels, instead of two legs, is something that comes up a lot in this lab, and in others. Researchers assume that people expect biped robots, like C-3PO in *Star Wars.* However, I am told repeatedly, these are in fact extremely difficult to build. While Boston Dynamics have distinguished themselves with the extraordinary feats of agility in their biped and quadruped robots, the team tell me, that is because they have the resources to specialize in this one specific area. For everyone else, including well-known robots like Softbank's social robot, Pepper, wheels are a much more stable and cost-effective option.

Stevie's face is designed in line with state-of-the-art HRI research, to maximize 'affect' and expressive capacity. His head and face were designed in collaboration with the National College for Art and Design in Dublin. When his face is 'on', it consists of two digital displays: an upper display shows two big brown eyes and eyebrows. A 3D effect using graphical software allows for a glint in the pupils. The lower display is a simple line drawing of a mouth. Even when Stevie is unengaged, his eyes appear to blink continually, his eyebrows move up and down, his head moves slightly from side to side and his arms perform irregular micro-movements. Together, these give the impression of the robot being animate. The two separate screens are based on literature from psychology suggesting that people perceive the top and the bottom of the head differently and, the team tell me, they are the first robotics team to implement it in this way.

The first Stevie prototype was covered in a soft, velvety, cream material, chosen so as not to appear too 'product-like' and inspired to some degree by the robot PARO. While at the lab, I participate in user trials, including one for selecting the 'optimum' material for the skin. This involves rating a range of materials according to my preference, both visually and touching them while blindfolded. Ultimately, the final material and colour are chosen in response to feedback from carers, who want something that is easy to clean and clinical. It needs to conform to medical standards suitable for care environments and is thus made of medical-grade anti-microbial material. It is shiny and white.

Stevie is covered in sensors, about sixty of them, including a LIDAR, or light-detection, sensor for measuring distance. There is a microphone for capturing audio data, webcams on the front and back to capture images and another webcam on the neck area with which to capture 3D. This will allow for depth perception, which a normal webcam, just capturing pixels and colour, cannot do. Proprioceptive sensors sense and measure Stevie's inner state, such as his position in space, which way he is facing, his speed and the angle of his joints.

The robot has the capacity to store the data that is captured locally and is also connected to the wider network of connected devices in the lab, as well as the internet and the cloud. An on-board wireless router is installed to enable off-board processing, teleoperation and remote data logging. This networked structure allows for all sorts of configurations and connections, such as distributed and 'off-board' processing, data storage and use in the cloud, as well as remote-controlled teleoperation. A person can take over the system and remotely operate Stevie's actions and speech using a laptop. He can also be used, like any 'smart' device such as a smartphone or tablet, as a communication device, such as for video or audio communication.

What might be referred to as Stevie's 'intelligence': his voice, his responses, his gestures and movements, are implemented using a few different approaches. A limited, predefined set of social interactions are programmed in. Stevie's speech recognition capability is implemented using IBM Watson's speech recognition API (application programming interface). Stevie talks using a speech synthesis SDK developed by Cereproc, an 'off-the-shelf' male-voice solution with a 'Southern English' accent, chosen after conducting trials with potential users. Stevie 'verbally communicates' through speakers in his ears. Stevie's autonomous behaviours are implemented using a common framework for writing robot software, called the Robot Operating System, or ROS. This includes tools, libraries and conventions to standardize and simplify the task of creating various robot behaviours, such as the 'Navigation Stack', which takes in information from the sensors and translates it into velocity commands that are then sent to the actuators. The team are also actively working on instituting machine-learning techniques in order to develop Stevie's 'social' abilities. In order for Stevie to 'learn' the dynamic and continuous flow of natural conversation, as well as how to recognize people, the system will need to be continuously supplied with a huge dataset of natural language, and 'trained' using examples.

In the first version of 'Stevie', computing was done with a single computer. The latest version has a distributed network of computers, which means there is no single central processor doing all of the processing, which slowed the system down, even sometimes incapacitating it. As well as an Intel NUC (i5 processor, 8GB RAM) on-board computer, there are a series of distributed lower-power computers and small motherboards, allowing for the processing to be done in parallel. Each set of cameras has its own dedicated computer just to process the visual data. Other smaller computing devices can be added into the wider system, such as mobile devices and microcontrollers. These distributed microcontrollers communicate with the main on-board computer using UART over USB. This approach is called 'AI on the edge', meaning the data processing and decisions are made locally, rather than in the centre, and allows for much faster response times. It also has the added benefit of being more secure and reliable. The main computers may be updating every second minute, but the smaller ones can update in milliseconds.

Stevie's existence owes as much to science fiction, people's responses and the creativity of his creators as it does to science. The robot is a concrete and enduring embodiment of specific, historical and local materials, practices and ideas, which are stabilized and endure in a way that concepts cannot. His form, like that of other social robots, is inspired by both *Star Wars* and Astro Boy. His stature and colour owe much to Japanese design aesthetics. Stevie's capabilities are both facilitated, and constrained, by the current state of available materials (metal, motors, plastic), mechanical engineering, battery power, digital and AI capabilities, availability of funding and researcher talent. He is also the result of thousands of small design decisions by the researchers working on him. The robot's identity as a single entity, as 'Stevie', is created by adding a face. It is not until it is encountered that the robot is identified as a unified object.

In the lab, the work is rarely on the whole robot. Most of the time, an older prototype of 'Stevie' and the new model are standing immobile and switched off, out of the way. The only time I see Stevie 'whole', is when he is about to be tested with potential users. Different parts of what will eventually become Stevie are laid out on tables and counters around the room, including microphones, webcams, sonar sensors and batteries.

> The 'robot', such as it is now, does not look at all like a robot. If I did not know, and had to guess, I would have described it as a transparent plastic coffee table, with several shelves and bits of wire and plastic sticking out of it. Possibly to remind people what they are aiming for here, there is a piece of wood with a face drawn on it in marker pen pinned on top. (Field notes, 23 October 2018)

Donal is assembling the robot parts. He is not a graduate student like the others in the lab; instead, he has taken a year out of his undergraduate studies to do this work. He is uncharacteristically humble and philosophical about his contribution to the lab. He primarily works on the electronics. In the interview, he describes Stevie's architecture in terms of 'hierarchies' and tells me he is working on the lower level:

> The lowest level is the mechanical system and on top of that I would have provisioned, installed and designed for batteries and the power supply. That's the most basic, you're not going to do anything without energy. Once you have that structure, that system, you then build on the most basic components. So: sensors. As you read what position each joint is and you have little, very basic computers which will do, not a lot, but will read what the joints are and then have some sort of idea about what the joints should be doing and then try to adjust for that. So that's a sort of cause and effect there. But in terms the overall goal of what the machine wants to do coherently, as a holistic complete level that doesn't really have much to do with the work I've done. So basically, I'm like a slave carrying out orders from the masters. Yeh, in the computer science terminology there's like slave devices, master devices and things like that. (Interview, 1 February 2019)

The team are adamant that they do not want the robot to be too human-like or to risk falling into the territory of the 'uncanny valley'.[2] What is more, they say, studies have shown that people's expectations are set by the outward appearance of the robot, and the team do not want anyone expecting too much human-like intelligence from Stevie. As is also observed by Kathleen Richardson (2015) in her ethnography of robotics researchers at MIT, the team at Trinity are very aware of critiques of the racial underpinning of robotics research. They are therefore somewhat apologetic of having produced such an extremely 'white' robot, and go to great lengths to explain the rationale (which is that it is for medical use). There are also accents of yellow, chosen for aesthetic reasons.

They have built a device for making the robot's 'skin'. It is a wooden contraption with two large shelves. The bottom shelf is waist high and has a wooden mould of an upper torso on it. Underneath this shelf there is a vacuum mechanism. On the underside of the top shelf, they have placed a plastic sheet. Then, in the same way as table ice hockey but in reverse, Daniel explains, the plastic sheet is sucked down and is moulded to the wooden block. They have four of these plastic skin pieces lying on the desk.

The researchers did not intend for Stevie to be identifiable as male or female. Stevie's name and appearance, they say, were chosen to be deliberately androgynous. Although efforts were made to implement a genderless voice, they tell me that users generally found this difficult to comprehend. After a period of testing involving users, one of the male voices received the best feedback and was implemented. Despite their aspiration to gender neutrality, then, the male voice clearly identifies the robot as male, and pretty much everyone refers to Stevie as 'he'.

Everyone, except the team, is very interested in Stevie's name. It is the first thing I ask them about. In all the time I spend with the team, it is also the first question that journalists, interviewers and potential funders ask. The team, on the other hand, are embarrassed by the question. It has so little to do with the skill and work that they feel that they bring to bear on the robot. The name is irrelevant, and incidental, they insist. Nonetheless, we persist. They tell me that Stevie's name originally resulted from a demand from those interacting with him. The team originally chose what they hoped was the gender-neutral 'Stevie' and subsequent attempts to rename him have been roundly rejected by a committed community of dedicated Stevie-philes. One of the engineers is particularly irritated by my question: 'The name is the least significant thing that we are working on ... you know? Call it whatever you want!' (Interview, 1 February 2019). But for the rest of us, once the face goes on the robot, its identity as a whole, as a Being, becomes primary.

Initially, the team thought that Stevie could be used to solve distinct problems, like attending to specific care needs. They had assumed that the next steps would be to put functioning arms on it and get it to do service tasks. Apart from the fact that these tasks are remarkably complex to replicate mechanically, initial feedback from potential users convinced them to focus on Stevie's communication and entertainment capabilities. For this reason, Stevie's arms are there purely to give him a human-like appearance and to carry out 'non-verbal' communication. They are deliberately designed not to let people think that they are useful and so end in small balls, rather than hands. He can, and does, use them for gesture, to convey emotion and direct attention.

Everyone in the lab agrees that his 'intelligence' is basic. This, they characterize as being 'introspective', knowing how to orient and manoeuvre itself in space, being able to talk, a budding ability to recognize and distinguish things in the world, with rudimentary interactive abilities.

> It's a very basic form of intelligence. But it exists and it exists distributed so if you pull his arm he'll fight back. And if you hold him longer, he'll fight back more. The dialogue system would have a certain amount of intelligence ... from a technical point of view the actual recognition of

> the speech is quite intelligent but we can't lay claim to that because that is sent on to Google ... parsing the language isn't really intelligent. (Interview, 24 January 2019)

A more developed intelligence, I am told, would feature greater autonomy, such as more autonomous movement, and an improved ability to react to people and things in the world, such as the ability to react to people's emotions:

> Well, when you say 'intelligent', especially in the context of machines, you would think of artificial intelligence as in human-level intelligence ... things like language, social interaction, understanding context, empathy, and also abstraction, reasoning ... if they could ponder, if they could create art and poetry.
>
> [Teleoperation] gets around the problem that AI isn't intelligent. Which it's not, not really. A robot cannot exist on its own in the world. Even the state-of-the-art. It will get stuck, things will go wrong, and it won't know what to do. (Interview, 24 January 2019)

Stevie's development is not funded by dedicated departmental or robotics funding; instead the team have managed to secure small amounts of funding in disparate areas such as innovation and healthcare. Ultimately, they hope to secure investors and develop a spin-off company. They are not part of a wider national robotics effort and, despite the engineering legacy of the University, there is little robotics history or institutional knowledge in this particular area. Despite this, or perhaps because of this, they are the only team that I meet who are attempting to develop a complete humanoid robot.

In a manner that is not at all typical for robotics teams, the team building Stevie have been engaging directly with communities of older people since the very beginning. During the time of my engagement with the lab, they have ongoing engagements with two pilot sites. The first is a charity in Dublin that provides support services for older people in need, and the second a retirement community in the east coast of the United States. The work in the lab is greatly intensified when preparing for these activities. When I arrive at the lab, before the first deployment in the US, the team are all furiously working on the new robot, emphatically not 'Stevie'. They are currently calling it 'Boole'; 'we are crap at names', they tell me. The new robot is much further along than the last time I had seen it. All the levels in the 'coffee table' are filling up. There are batteries at the bottom, cameras and sensors up on top. There is what looks like a little fan in the middle. Not-Stevie's face has had a makeover. His eyes are bright blue eyes, with a 3D-rendered glint. Everyone looked exhausted. The robot is full up with wires and mini-motherboards and sensors and things I do not recognize. There are two wheels at the side and a ball at the back. The front, unlike the original Stevie, has a platform, which someone is trying to pin an iPad to. There is a blue light shining out of the bottom of the robot which looks quite cool. Donal is programming in something to do with the robot's movement. Not-Stevie is now moving around quite smoothly. There is less noise than the original, which is a huge improvement. I ask what the robot will do when they go to the States. They are working on two kinds of functionality. First, they want to allow someone, a carer, to teleoperate it using mobile goggles. Second, they want it to facilitate bingo for the community.

When I return, after the trial in the US, everything is quiet again in the lab. The robot is called Stevie again and its eyes are now brown. It is standing in the middle of the room looking rather dejected, arms hanging down and head to the side. The upper body is separated from the lower third and I can see a bunch of wires going up the middle, like a mechanical spinal cord. The back of the lower half is also open and exposed, and connected to a very large battery. It does not seem like Stevie has been all over the world and back again, impressing all the old folks. But he has, and he did. They are planning a second trial in August, and this time I will be going with them.

Over the next couple of weeks Stevie starts to look a lot better. The lab is tidier. There are fewer people around, and those that are, are calmly working at their computers. An impressive, sleeker Stevie stands in the centre of the room, with gleaming yellow lights, adding panache. His physical structure is 'rounder'; more womanly they tell me. He no longer has broad shoulders and narrow hips. This was necessary to allow for extra space, but also the team are still trying to resist any definitive gender classification. This is tricky, as no one has ever thought of Stevie as anything other than a 'he'. His design, Daniel tells me, has emerged both from the technological constraints, feedback from those interacting with him, as well as the team's creativity. They build on each other's ideas, seeing what works and what does not.

Heriot-Watt, Edinburgh

In Heriot-Watt University in Edinburgh, Scotland, Gerry and his team are working on social robots. I first travelled there in 2018 to carry out fieldwork and returned the following year for more fieldwork and interviews. The robotics labs are housed at the School of Mathematical and Computer Sciences, with labs dedicated to swarm robotics, drones and HRI. During both of my visits, I am given a desk with a number of PhD students, but most of the days I spend following Gerry around, observing and discussing his work. Much of his day is spent attending project meetings discussing current and future projects. Unlike the team working on Stevie, the teams in Edinburgh are not building their own robots; rather, the bulk of their work is focused on understanding and implementing novel approaches to human interaction with many different robot types. The HRI lab is a very large room filled all sorts of humanoid and zoomorphic robots. A large 'FLASH' robot stands in the corner going through a range of facial movements or emotions. Opposite, an adorable iCub is going through a yoga routine. Under a table is Sony's robot dog, AIBO. On the table, there are two small robots, a Miro, a 'bio-mimetic' robot designed by a collaboration between UK roboticists and the Sebastian Conran's design studio. On the table were a few Anki Cosmo robots, mostly used for outreach and teaching.

Gerry's work is focused on combining machine-learning techniques with computational decision-making programs. The project he is working on at the time of my visit is one in which they are manipulating the robot's expressive capabilities to see how people respond to them. Like Stevie, this robot's head has only three degrees of freedom, so they have to test whether the expressions they designed are still interpreted in the way that they want them to be by humans. Another researcher, Lena, is working on 'social signal processing', which is concerned with modelling observable social cues that are exhibited by people in social interactions. She is looking specifically at the problem of the robot knowing when to interrupt a person at the appropriate moment, a behaviour that, she tells me, is particularly important in hazardous environments. The method she is using is to try to detect 'cognitive' or 'mental load' in the user. This they are doing by measuring pupil diameter, which has, Lena tells me, been used in cars to see if the driver is still attentive. Thus, the system might recognize the pupil diameter and then interrupt, or not interrupt, on that basis, or change the behaviour of the robot system. They are also measuring heart-rate variability, skin conductants and blinking frequency.

Sara, a professor in the department, is focused on models of 'action selection', and questions such as: 'How does the robot know what to do?' AI, according to her, is the art of 'doing the right thing'. The set of possibilities is finite, she explains, constrained by the physical nature of the artefact. This then needs to be engineered on a number of different layers of abstraction:

> I can go to the office today. That would be a high level of abstraction. I can walk towards my front door and turn the knob and that would be a much lower level of abstraction although quite closely related because I might be doing that so I can go to the office. So, another abstraction. Time scales. Abstract actions have long time scales. Concrete actions have short time scales. You shouldn't be so focused on getting to the office that when a fire breaks out in the kitchen behind you, you still go to the office. That would be silly. We're good at that type of thing. We don't persist with goals when it is clear the goals should be substituted with something else, because we manage our goals quite flexibly, which is much harder for robots to do. (Interview, 27 June 2019)

Sara is using models of empathy and affect to try to model these behaviours. Separate to these, she tells me, is social context, which can be brought to bear through the use of techniques such as roles, narrative and games: 'A role will give you a kind of a determining envelope for behaviours. If they are the boss and you are the subordinate. You have a social relationship. This is all culturally mediated. Power relationships are culturally mediated. Gender has an influence and that depends on the culture as well' (Interview, 27 June 2019). A combination of the three: empathy, affect and social context, she says, will allow them not just to reflect the internal-model affective state of the robot, but also to project the consequence of a particular expressive behaviour as 'an action'. This will allow the robot to reflect on, and judge, whether a particular action it carried out was appropriate or not. However, she says, tracking social context is incredibly difficult, even having a vocabulary for what the social context is, and connecting that to the appropriate social signal. The solution, she says, would be to get 'theory of mind' working accurately. However, what this ultimately means, she admits, is not entirely clear: 'There isn't universal agreement in psychology about this. I'm only a computational person; I can only take what I think I can ... I take a view from psychology; it's not the only view' (Interview, 27 June 2019).

However, as we have seen, in actual experiments with robots in the field, or 'in the wild' as it is commonly referred to by roboticists, it tends to be much more prosaic issues that present the biggest challenge. In this case, it was the issue of ensuring the robot had enough power to keep running:

> We ran a robot for three weeks continuously but we made a serious design error, which should have been obvious but wasn't, like other serious design errors. So like I say, robots can run for about two hours without needing to recharge, so we designed in an automatic recharge behaviour for the robot ... we had the recharging unit in the corner and whenever its battery was low the robot would drive into the recharging station and recharge. The easy way of doing this is to drive in frontwards. Big mistake! Well, it can't interact with anybody, can it? It's facing the wall. We should have thought of that! And people got really upset about it, because, you know, the robot would have to recharge for two or three hours. It takes time to recharge a robot. It doesn't happen just like that. They are big pieces of machinery, typically, with big batteries in them that aren't going to recharge in just five minutes. So, it takes two or three hours to charge the robot as well. And during that time, it wasn't able to interact. Had we had it reversed while in the charging station that would have gone better. (Interview, 27 June 2019)

Across campus, a Living Lab has been set up as a model apartment complete with kitchen, sitting room and bedroom with double bed and wardrobe, and an en-suite bathroom. Lucas shows me around the lab that is intended to test the robots as part of a wider 'smart' environment within a 'natural' setting. There are two robots in the apartment, a Pepper, focused on social interaction, and the vaguely humanoid PAL TIAGo, designed to perform tasks, like picking up objects. The floor is smooth linoleum throughout and there are no tricky areas like awkward lips between floors. There are sensors under the floorboards and in the kitchen cupboards, and cameras in the corners. Later, when I interview Lucas, he is eager to stress the pragmatic nature of his work. He is not interested, he stresses, in understanding 'the inner mechanisms of intelligence in humans', instead he's trying to 'emulate it with planning, classical AI combined with machine learning' (Interview, 26 June 2019).

Towards the end, the interview becomes more animated. Lucas tells me about an article that he has just read in the UK *Times* about a forthcoming book by James Lovelock, creator of the Gaia hypothesis (Gill 2019). According to Lovelock, Novacene, or the age of cyborgs, will supersede the Anthropocene. For Lovelock, this represents the next stage in human evolution. These cyborgs will not be humanoid, an idea about which Lovelock is scornful, according to the article. Rather, they will likely be spheres. But we do not need to worry about them killing us, because these new beings will need us just as we need plants. As part of Gaia, it will be in these new cyborgs' interest to continue to regulate the planet's ecosystem and keep us cool (below fifty degrees Celsius). This future is driven by evolution and natural selection, says Lovelock, 'you can't avoid it'. Lovelock is an eco-futurist, believing that engineering will save humans from catastrophic and existential harm. 'It's a realistic picture, I think,' Lucas says.

A huge amount of researchers' time in Heriot-Watt is dedicated to doing demos for schools, for the public and for the media. Their relationship with the media is fraught, they tell me, having had a particularly damaging experience when a week-long experiment of embedding a robot in a local clothes shop had resulted in negative global coverage of a robot being 'fired'. With a few exceptions, when asked, most of the researchers that I interview express their frustration with how media report on robots, how they inflate expectations and scaremonger, and how they misrepresent the research. There were very few who enjoyed an easy relationship with the media, although, as we shall see, some benefitted considerably from media attention. And yet, while it is undoubtedly true that there is a strong public appetite for stories of imminent AGI in humanoid dressing, it is not the case that these stories are solely being perpetuated by media or marketing rhetoric. There is a requirement on the part of researchers to maintain the mystique of imminent AI to ensure that the interest, and the funding, are maintained. 'What tends to happen is that people over promise, under pressure, because you don't get the money if you don't and then people on the other end who have far too high expectations say you aren't delivering, are you? This is crap, I wanted something else. Crash! You know? I've been here before' (Interview, 27 June 2019).

Bristol Robotics Lab, Bristol

Bristol Robotics Lab (BRL) in the south-west of England is a dedicated robotics space, the largest in the UK and unlike any other lab that I had visited up to that point. It is an impressively large building surrounded by artfully managed flowers and structural grasses. Two or three massive industrial robotics arms function as spotlights at the front. There are several portable toilets stationed outside, which I later discover are an innovation that has arisen from research into biofuels to power the robots. As a result, as well as robots, they also develop toilets with electric lighting, for use by women in India to increase their safety, and at the Glastonbury music festival. There is a reception area with a sign telling us that photography is forbidden. I have to wait here until someone comes to meet me. Inside there are open-plan, yet intimate, spaces bounded not-too-linearly by glass walls that double as whiteboards covered in impressively complicated-looking algorithms. Each cube hosts a different research area, such as swarm robotics, social robots, 'soft robotics', tactile sensing robots, robot prosthetics. In the centre of the space is a large aerodrome for drones and aerial robots.

A substantial research focus at BRL is devoted to the problem of energy and the amount of it needed for the robot to 'maintain autonomy'. The bioenergy and self-sustaining robot research groups are focused on 'feeding' the robots by generating electricity from naturally occurring biomass. The three sources used are urine, flies and slugs. A small robot machine outside the door on a set of tracks is powered by flies, acting a bit like a Venus flytrap plant might, slowly dissolving the flies in its mechanism. The organic waste is converted into electricity using 'microbial fuel cells'. Like in biology departments, the doors to these labs are closed and display hazard signs. In 2013, the team at BRL developed a way to charge mobile phones using neat urine (Ieropoulos et al. 2013). In the toilets at the lab, cardboard bowls are provided so that researchers may harvest their own biofuel on site.

Richard, whom I first met at AISB, is my key interlocutor here and I have been given a desk with his team. He is one of the founders of the lab and no longer works there full-time. Richard's interest is in programming robots to engage in ethical behaviour, and his students are coming at the problem in disparate ways. One of them, Danni, is working on a project with swarm robots focused on emulating group behaviour and 'collective decision making'. Swarm robots are small and simple physical robots out of which, working together, more collective and complex behaviour 'emerges'. They can be programmed to react to each other in simple ways; for example, 'cooperation' mode might describe a number of robots programmed to focus on the same task. In 'organization' mode, a single robot may control the others. Behind us, there is a white plastic surface with one thousand tiny 'kilo-bots'. When they are on, their simple sense-and-act system causes them to move towards, and then be repelled by, a neighbouring robot so as not to hit it. The result is a remarkably lifelike display, which, in combination with the sheer number of robots, sounds like heavy rain on the plastic roof. It feels like a swarm, but because the robots are tiny and shiny and silver with blinking lights, it feels a lot less repellent than an insect or rodent swarm might. They also have larger swarm robots, called 'e-pucks', which are 'smarter', that is, they have more computational power and can be programmed to do more. In Danni's study, she is programming the robots to collaborate on moving a frisbee across a platform. First, it is programmed in a simulation. The robots have to 'adapt and learn', using programmed positive and negative feedback to build 'consensus'. Danni talks about a scenario where these robots could be used, for example, in old, abandoned nuclear reactors where you could just cut a small hole in the wall and let these robots in. The robots, she says, unlike humans, are expendable. So, while some of them might 'die' from the nuclear toxins, enough of them could survive and do whatever job needed to be done.

Another student, Max, is working on developing an internal model for a robot, in which a robot runs a simulation of itself and its environment inside itself. On the screen, Max has created a 3D simulation in which the robot is represented by a cube. The small robot-cube is positioned on top of another, larger cube, which represents the ground. There are other potential obstacles, represented by different 3D shapes: a cone, a triangle and a cylinder. Ultimately, this will allow a robot to operate as a 'real-time consequence engine', allowing it to model and predict the consequences of its own actions, and those of other agents in the environment. It is hoped that this might be a step towards building self-aware behaviour, and ultimately even ethical behaviour. At the moment, for example, a robot might be able to make ethical decisions such as stopping other agents from falling down a hole.

Between the research labs and the restaurant, there is an enclosed space, which Richard tells me is the 'incubator'. It is not, I am disappointed to find out, full of 'baby' robots but, rather, a business incubator where people who have start-ups in robot-related areas can rent some space. I tell Richard about the baby robots. He is not as amused as I expected. He explains that they do, in fact, have a plan for 'evolutionary' robots, in which the robot will build itself in a 3D printer, and then emerge into a robot-learning environment, where it can learn from other robots (but only those robots that are themselves sufficiently advanced). Only the most successful robot will go on to 'reproduce' their algorithms in the next generation.

At one end of the lab there is a display of 'dead' robots, including an early Shrewbot. Shrewbot is particular to the lab at Bristol and consists of a body with a single mechanical arm out of which whisker-like tactile sensors protrude. It does not have to rely on light for vision as other robots do, and so can navigate down pipes and in the dark. Most of the robots at the lab in Bristol are not humanoid. Humanoid robots feature in just two research areas, a small living lab in the centre of the space, and the HRI research group.

After presenting my own research to the group at the lab, I am adopted by two HRI researchers, Lucy and Stefan. Lucy is a PhD student and is looking at the potential for robots to influence human behaviour. In her study, three robots try to persuade someone that they should exercise. The first robot is programmed to use 'non-committal technical language', the second will use the first-person pronoun 'I', and the final robot will quote others, such as 'your therapist told me to tell you …' The test will evaluate which type of prompt is more persuasive. Lucy and Stefan are excited about the potential for machine learning to create 'social behavioural datasets'. Lucy compares machine learning to grounded theory and wonders whether it could beat an anthropologist in identifying phenomena in data.

Like all of the other labs, my presence in here is not in the least noteworthy for teams who are used to a constant stream of commercial funders and journalists. Indeed, when I first arrive at any of the labs, including this one, I am generally offered a tour, in which researchers working on their robots dutifully pause, stand up and explain their research to me as if I am some visiting dignitary. My presence offers another opportunity for junior researchers to practise their demonstration and media communications skills, which is such a major part of this research, and something they are constantly in training for.

Carnegie-Mellon University, Pittsburgh

As we have seen, it was in the US, rather than in Europe, or even Japan, that the creation of intelligent robots first developed as a serious research programme. The 'big four' centres had been working in earnest on trying to create human intelligence since the first research grants in the early 1950s. According to McCorduck (2004: 133–34), each of the four main centres evolved its own style based on the personalities of the men who dominated it. While MIT (after Minsky), was 'haute couture', 'stylish yet faddish', Carnegie-Mellon University, or CMU (after Newell and Simon), was 'classic but stodgy', representing old-world artisanry, attending to detail and using the finest materials. Both Suchman (2007) and Richardson (2015) conducted fieldwork at CSAIL at MIT in the early 2000s. My research took me to another key centre, the robotics research institute at Carnegie Mellon in Pittsburgh, which I visited in October 2019, carrying out observation and interviews.

To a visitor unfamiliar with the geography of the US, there is an almost eerie juxtaposition between the high-tech architecture and prestige of the University and the rusty, industrial and shabbily-faded glory of the surrounding area. Pittsburgh is a proudly industrial town, known as the 'Steel city', having produced 60 per cent of US steel between 1870 and 1910. The legacy of this economic boom is still evident in the cultural centres, many of which are the legacy of industrialist Andrew Carnegie, including the University and the Pittsburgh museums. Westinghouse was based in Pittsburgh and Herbert Televox was first displayed here. Robots are part of the city's history and identity. However, there was a period of deindustrialization in the 1970s and 1980s, which brought decline, and with it the loss of 130,000 manufacturing jobs, mostly in steel. Unemployment was at 18 per cent (Andes et al. 2017). Now, along with states such as Ohio and Indiana, Pittsburgh is known as part of the 'rust belt'.

When I first arrived in one of Pittsburgh's less salubrious suburbs for my stay, the decay was palpable. Sidewalks and services were crumbling, metal signs and bridges were rusted, electric wires hung low across roads and there was a deeply unpleasant odour pervading the air. Today, Pittsburghians' relationship with technology is fraught. While the population benefits from the dwindling iron and steel industry, it also pumps out harmful emissions (Maher 2019). A report from the National Bureau of Economic Research (NBER) warns of concerns over Pittsburgh's high 'exposure' to robots (Acemoglu and Restrepo 2017).

During my visit, Google Maps proves to be useless, as the fact that many of the streets weave over and under each other on different levels is lost to the Uber App and, as a result, also on the Uber drivers. Thus, my first impression of the university is not the impressive Beaux-Arts style buildings and wide promenade that greet the visitor to the main entrance of the University at Forbes Avenue but, rather, a service entrance at the back of a car park underneath one of the side buildings.

The Robotics Institute is part of the School of Computer Science and, founded in 1979, one of the oldest robotics departments in the world. It is huge, with over fifty full-time faculty members, spanning seven floors and several buildings. Bob, a veteran CMU roboticist, agrees to show me around. The newest building is the Newell-Simon Hall, which was built, Bob tells me, to rival MIT's Frank Gehry building. He points out a photo of Newell and Simon in the entrance hall, 'we each have our founding fathers', he says, referencing MIT and Stanford; 'these', he says proudly, 'are ours'.

Opposite is the boarded-up cubicle that used to house the 'roboceptionist', a collaboration between the Robotics Institute and the School of Drama. The original character, 'Valerie', with a 'brilliant blue-eyed gaze', would sense people walking past and offer them assistance (Gutkind 2009: 104–6). Most of the time, however, Valerie spent on the phone revealing secret details of her private life, intimate conversations with her psychiatrist and aspirations to be a lounge singer. Perhaps it was the breathtakingly gendered stereotyping that led Valerie to be replaced, in 2005, by 'Tank', 'son of a NASA scientist who had aspirations to be an astronaut but was grounded by his vertigo'. By the time I arrived, the roboceptionist was gone. No one could tell me where.

The idea of robots roaming the halls of the academic centres that birthed them has become almost a cliché, reported endlessly by journalists, academics and other visitors to the centres. And so, I was a little disappointed not to encounter any. There were a few static ones, however. Up on the first floor, an immobile robot senses passers-by and tries to engage them in conversation. In the restaurant, a robot chess player, the image now decentred in the tilted TV screen, challenges people to a game of chess. None of the students or faculty at CMU showed even the slightest interest in these displays. They are, however, shown to visitors, who obligingly and inevitably attempt short, awkward conversations with them.

The HRI institute at CMU is part of a separate, dedicated human–computer interaction and design department, which although part of the computer science department, is not part of the Robotics Institute. It is, however, also housed in Newell-Simon. According to Bob, it is the 'left-side' of the brain to the Robotics Institute's 'right-side'. Although I am based on the Robotics Institute, I do visit the HRI department. Inside, the walls are colourfully painted, and feature lots of sofas and 'break out' areas. Upstairs one of the workshops has displays on the walls of 'smart materials' for sensing and moving. There are beautiful, knitted robots, 3D-printed plastic objects, foam and cardboard models. A large room dedicated to 'futures work' allows for the prototyping of environments, such as, one of the researchers tells me, a completely smart home for teenagers, which they used to understand how much 'smart' technology was too much. They also used it to recreate 'service' environments, so they might build a shop, or a restaurant, out of cardboard. The futures work, I am told, allows them to experiment with possible technologies without having to have those technologies in place yet.

Each of the different research areas of the Robotics Institute is spread across the six floors of Newell-Simon and parts of several other buildings. Newell-Simon is built on top of the ruin of the old United

States Bureau of Mines. There are still train tracks coming into one of the lower floors. We see a pair of very old 'Adept' robots that Bob says he has rescued from storage. They are huge, very solid machines, with a single arm covered in extendable plastic. This is the challenge for soft robotics, he says, as the rubber on one of them has all but disintegrated.

Across the many floors, there are rooms with different research areas. 'Bio-robotics' is at the top. The 'bio' in the title is loosely interpreted, meaning they look to biology for inspiration. In this case, it appears to mean based on animal physiology, rather than attempting to create a humanoid. The robots here are modular and can be put together in any configuration. The first thing I see is a robot snake. There is also a snake obstacle course, and a snake sandpit. Another robot made of the same modular sections but assembled in a different way resembles a spider, with eight legs.

In one large room, a researcher is listening to 'Scarborough Fair' by Simon and Garfunkel and growing plants in 'autonomous environments'. In the lower levels, there is a massive arena with a huge water tank for underwater robotics, and a netted area for testing drones. Here I can also see industrial robots, self-driving vehicles, lunar robots, arctic explorers and enormous spider robots. At the sides of the arena there are 'workshops' full of people making things. Here they are referred to as 'shops' rather than labs. Another level down is 'the catacombs', which are using for testing subterranean robots. This part is left over from the original buildings. Some parts are not even head height. There is a massive fan and an enormous block of coal.

There is a 'panoptic studio': a huge, domed, igloo-shaped structure built to capture motion and social interactions in 3D. The system has 480 synchronized video streams so that it can capture a large amount of people's movement in space. This then produces a 3D output of 'labelled, time-varying anatomical landmarks'. These movements, captured algorithmically, can be used both to predict and to reconstruct 'skeletal trajectories'. There is a manipulation lab, and a lab for connecting vision and manipulation in which a gripper is trying to pick up coffee beans.

Manipulation is a big part of the robotics research at CMU, and one more associated with industrial robots than social ones. According to Nick, a roboticist at CMU since the 1980s, manipulation, or 'making other things move', was originally what robotics was about, and people were more interested in manipulation than in the mobility and locomotion of a robot itself. I interview Nick in his office. Behind him, there are lots of medals and awards, including some from DARPA. Back in the 1980s, he tells me, robots were almost entirely destined for manufacturing. 'And by the way', he tells me, 'it's also not clear that those industrial robots that I talked about should be called robots. That was a subject of dispute. Maybe still is in some quarters. The reason they are called robots is because of Joel Engleberger, the guy who founded Unimation. He thought it was a good marketing thing to call them a robot' (Interview, 9 October 2019). Like many roboticists, Nick is also a chief scientist in a company developing robotics for industry. For Nick, the ultimate test of a robotics device is to succeed commercially, to find a niche where it can do something that is economical relative to human beings.

Nick is defensive of people who consider manipulation uninteresting or accuse it of not having to deal with uncertainty. For Nick, making an instrument that deals with precision is always about dealing with uncertainty, and thus also about intelligence. He talks about mechanical intelligence and morphological intelligence, and the fact that intelligence can 'drift back and forth' between the mechanics or hardware, the software and the designer.

Tina is also a professor at the manipulation lab. She was drawn to manipulation, she tells me, because of the precision and the 'science' of it, rather than what she calls the 'tinkering' of robotics. Tina's team are working on soft robot hands, using materials like rubber, cloth and foam. The hands, which are coloured green and purple, have four fingers and wobble in every direction, looking eerily alien. One of the advantages of soft fingers, Tina tells me, is their ability to collide with each other without incident. 'With a rigid robot we're always worried: oh, it's going to collide with something, it's going to break! And there's all this concern about something so simple as the fingers running into each other. And with the soft hand it's just such a relief we don't care if they run into each other. Nothing bad is going to happen' (Interview, 15 October 2019). Tina would like to develop a 'grand unified theory of grasping and manipulation', a classification system that would describe how people grasp and manipulate objects. She suspects that there are a finite number of things, perhaps a hundred, that a robot could do that would render them human-level capable. It is just about finding which hundred, she says.

I also interview Bob in his office. He has been at CMU for nearly twenty years. Before coming here, he held positions at other prestigious institutions. He tells me he is sceptical of other researchers (which includes Breazeal to most of my informants) who trace their interest in robots back to R2-D2 and C-3PO in *Star Wars*:

> That's just a myth, a story that they have created to explain to themselves a much longer term and deeper phenomenon. If the ground is

> not prepared, *Star Wars* isn't going to do it for you. In my case, I was a very clumsy, chubby, incompetent child and I had many brothers and my sister really liked to point it out ... so the idea of trying to understand that and master it was very attractive to me. And that is why I am doing what I do. (Interview, 8 October 2019)

Unlike many of my other informants, Bob is not concerned by media representation of his field. In fact, he is the second of my informants to have a close relationship with Hollywood, having worked with major studios on movie productions featuring robots.

Bob's office is full of junk. It is like a mechanic's workshop on top of which someone has dumped a whole load of ageing computing machinery from on high. He has to climb over things to get to the desk. For Bob, the architecture, and the junk, is integral: 'I think the architecture of a building really has an influence on how people think, which is why I like junk, it helps me think. You're not anybody without a pile of trash. Problem is, they keep tidying it up' (Interview, 8 October 2019).

While Bob has raised chaotic junk to an art form, just like at all of the other labs that I visit, at CMU there is robot and machine detritus everywhere. Robot hands lie on tables, robots stand motionless facing walls, and bits of robots are thrown into corners, having been raided for parts. Tables and shelves are covered with plastic boxes full of wires, nuts and bolts, labels, screws, paper, bits of plastic and foam, electrical circuits, old motherboards, wood, connectors, papers with diagrams. Some labs are more deliberate with this aesthetic than others. In both BRL and CMU, there are semi-formal 'roboseums', featuring select early prototypes and experiments that have been retired, and have gained a retro feel. The sight of abandoned bits of robots is, for the roboticists, simply characteristic of a workshop. To an outsider, it is somewhat unsettling. Behind a whole load of displays, I spot a pair of eyes:

> What's that? I ask. That's our man machine, he says. It's been abandoned. This is where robots go when they are retired. Man machine stares out at nothing without eyelids. What do you call him? I ask. It's a humanoid? He offers. There is no sentimentality here. On the table across from this is an exoskeleton attached to a pair of large, tan boots. This will make you walk faster, I'm told. There is a large, wooden ball in the middle of the room. What's that? I ask. You could try and stand on it, he suggests, if you have good insurance. (Field notes, 16 October 2019)

Outside, in one of the warehouses between robotics buildings, there is a large sculpture of an eagle, which one of their robotics teams won for the 2007 DARPA urban challenge. Apart from the prestige that accompanies such a feat for CMU, the challenge also included a cash prize of $2 million. A team from CMU also shared the $3.5 million cash prize at the 2015 DARPA Robotics Challenge, using Boston Dynamics robots. While DARPA remains the dominant funder when it comes to robotics research in the US, the influence of industry and venture capital on robotics funding is growing. A new 'big four' have emerged: Facebook, Amazon, Google and Apple.[3] At CMU, Disney and Apple have spaces just outside the University campus, and, during my visit, Facebook are in the process of setting up there. During my time at CMU, I have been given a desk at the Community Robotics, Education, and Technology Empowerment (CREATE) lab, which, as it happens, is in the same building.

The research agenda at the CREATE lab is different than the rest of the Robotics Institute, where the research is primarily motivated by the military and industry. CREATE was founded with the explicit aim of refusing military funding. Their biggest challenge, one of the directors tells me, was, if not for the military, then what were their robots for?

> Then the question became now, what's the positive? You know what the negative is, what you are not doing. So, what are you doing? And indeed, that is the situation. If we are not taking defence funding, how do we get funding in a space where, back then, so much, or the majority of the work being done is defence funded? And the answer to us was education. (Interview, 11 October 2019)

Like the other robotics research spaces, this lab also has packed shelves, mechanical bits, breakout couches and abandoned robots. But there is a difference. There is a much wider array of objects and technology hybrids. I can see feathers and bells, wires and glass balls, and something that looks like a car radio from the 1980s. A bottle of wine could be a project or a present. There are two cardboard figures: one waves a flag, the other looks like an Egyptian puppet. There is a shoebox with pipe-cleaners coming out of it. There are a few small, wheeled robot cars with wires, platforms and cameras. There are also a large metal racks with over twenty servers stacked on them. This, I will learn, has become the central focus of the CREATE team's work. The researchers here are a mix of roboticists, programmers, educationalists, artists and researchers from other disciplines, like geography.

The focus of the lab is 'to turn things around'; that is, rather than trying to figure out how to get people to accept robots, they start with people and communities, and then try to figure out how the researchers' tools, expertise and prestige can be used to serve them. This means that the lab is deeply embedded in the local community from Pittsburgh and its surrounds. On the tables between two couches is a small robotics project designed for pre-schoolers, called 'MindfulNest'. The interface is a tablet with a bright green cover. Attached to it is a black box which acts as a holder for three items – a plastic black and white flower with a green handle, a white tube that is rounded at the top, and a long blue tube. If a child picks up and smells the flower slowly, a scent will come out. If they sniff too hard – nothing will. If they pump the white tube slowly, balloons will float up on the screen. If they use the blue tube to 'conduct' music that is playing gently and with flowing movements, music will play. If they move it back and forwards too fast, nothing will happen. In schools, kids had taken to going over to it when they themselves decided that they needed to calm down.

Barbara's background is in geography and mapping and she is working on the air quality project in the Monongahela Valley. The steelworks and coking plants that are releasing emissions are affecting the air quality and therefore also the health of the people at the intersection of the three rivers. There is also a large fracking industry. There appears to be a higher rate of cancer and certain illnesses that may be related to the pollution from these industries, but it is difficult to prove definitively. The CREATE team have identified three independent sources of air pollution and are tracking them. Working with families in the area, they are monitoring the air quality using robots that measure it periodically throughout the day and night. The project has found, for example, that at 3 a.m. the levels of benzene go up. However, as Barbara relates, the issue and the solutions are not straightforward. Despite the negative effects of industry, the steelworks and the fracking are a point of pride for people. It means they do not have to rely on foreign oil.

Another project focuses on mapping the local housing situation, such as house prices, where buyers are coming from, tracking gentrification in the Pittsburgh area. The project visualizes inequities in Pittsburgh, including social structural inequities around housing, around education, as well as around factors such as the education gap between different ethnic communities. Beth is giving a presentation to local agency workers. As well as the quantitative data that she presents on an interactive map of Pittsburgh, she also presents qualitative data: interviews with local residents and detailed, up-close photographs of different parts of the locality.

The team also provide spatial and temporal imaging from all around the world, looking at pressing societal issues such as poverty, violence, modern slavery and climate change. One of these projects, called

Earthtime, is shared with global leaders at the Davos World Economic Forum. At CREATE, they are using technology to empower people who would otherwise not be able to afford, or have access to, it. Funding for the CREATE lab is not as straightforward as for the rest of the Institute. It is a balancing act, one of the researchers tells me. Each project has an eclectic group of funding relationships. Heinz Endowments is a major source. The pro-social orientation of the work has made physical robots almost redundant at the lab, and certainly anything that might be humanoid. 'We take the word "robotics" very broadly ... We use it to mean any system that has sensing, and complex machinery for cognition, and some kind of pushing on the world or action' (Interview, 11 October 2019).

The robot in the CREATE lab is less of a physical object and more of a symbol and rallying point, bringing together the capital, resources, data, power and social influence of 'robotics'. Instead of using it to increase power and profit, it redirects it, in an uncharacteristic inversion, away from the 'hegemonic power structures, to challenge negotiations of power in society and to enact pro-social change' (Interview, 11 October 2019). The lab offers a small glimpse of how things could be different. However, I suspect that its existence on the periphery of the University proper is the result of the determination of a few dedicated individuals, making its ability to influence the wider institutional network of robotics research less likely. In the wider Robotics Institute, there is still the idea that politics is separate from the field. In a long and rather combative section of the interview, in which I try to ask whether he has any concerns with regard to the impact of robots, including the data they need to 'learn', an exasperated Bob eventually exclaims:

> Now you're asking me questions that are political philosophy! Holy crap! I don't know! You are asking me very deep political questions! You guys are confusing ... you think robots and AI have anything to do with these issues and they don't [bangs table for emphasis]! Not in any deep way! ... It's like asking should we have coal-burning power plants or nuclear power plants. It has something to do with robotics because robots run on electrons. But come on! Don't ask the roboticists about it. They don't have anything to say! (Interview, 8 October 2019)

The Space of Robotics Research

For philosopher and sociologist Henri Lefebvre (1991: 38–39), 'conceived space', or 'representations of space', is the abstract space as understood quantitatively. It the 'dominant space in any society' and is 'the space of scientists, planners, urbanists, technocratic subdividers and social engineers, as of a certain type of artist with a scientific bent – all of whom identify what is lived and what is perceived with what is conceived'. Although representations of space are abstract, 'their interventions occur by way of construction', such as architecture, as 'a project embedded in a spatial context and a texture which call for "representations" that will not vanish into the symbolic or imaginary realms' (Lefebvre 1991: 41–42). Crucially, 'conceived space is thought by those who make use of it to be true, despite the fact – or perhaps because of the fact – that it is geometrical: because it is a medium for objects, an object itself, and a locus of the objectification of plans' (Lefebvre 1991: 361). This is also the conception of space that dominates robotics work.

For roboticists, a core problem of space is translating from the extreme reduction of minimalist 3D simulated space to the 'real world', the physical world of texture, friction and relative unpredictability. Thus, the art and craft of robotics also sees the practitioner in embodied and reflective conversations with the material world. As Agre (1997: 40) has observed, this differentiates engineers from more abstract sciences, such as mathematics. Roboticists are actively engaged with the material and spatial work from which they get concrete feedback. It either works or it does not. Their work *feels* true. It is, as Agre (1997: 11) describes it, 'nothing less grand or more specific than an inquiry into physical realizations as such ... what truly founds computational work is the practitioner's evolving sense of what can be built and what cannot. This sense, at least on good days, is a glimpse of reality itself.'

The advent of HRI and social robotics takes roboticists out of the field of engineering and computer science, in which producing a novel object that 'works' is in itself sufficient to render it 'scientific' and 'true', into the murkier world of the human sciences. Thus, robotics researchers find the requirements of their field growing, absorbing all domains. Already swamped with the difficulties of moving from the virtual to the physical world, the 'social' world outside the lab means that 'the wild', for roboticists, is a genuinely baffling place. It is little wonder, then, as Richardson (2015) also observed, that for robotics scientists the 'social' is located in the micro-behavioural exchanges between human and robot. This was also true in my research, as well as the fact that experiments with social robots were dominated almost entirely by quantitative methods in laboratory settings.

Roboticists regularly come up against the high expectations of those who encounter their robots, people whose expectations are shaped by watching science fiction movies and TV programmes and find themselves rather underwhelmed by the reality of robot capabilities. The highly challenging integration of hardware, engineering and computational resources that is required to instantiate this humanoid, mobile, interactive entity goes unnoticed, as people react more strongly to the blinking of the robot's eyes and appear instinctively to search for limitations and shortcomings in the robot's apparent intelligence. As we have seen, the field of HRI was established to exploit this tendency. However, not all of those in robotics are impressed by this development. Nick is dismissive of some of those from other disciplines developing an interest in robots, especially 'crazy HRI types', including, he has decided, me.

> You say to somebody in HRI what would be the first thing you would do to make this robot more appealing, or more acceptable, or whatever. Their answer is always 'you should put eyes on it' ... and people say, you know, in HRI they like to talk about eyebrows. Eyebrows! The old hard-nosed crowd, well that's something that they can goof on ... You could ask the question why is it called robotics? Robotics meant to a group of people something more specific. Maybe it should be called robotics. I'm not saying it should or it shouldn't ... I just find it difficult to care that much about how humans are going to interact with machines when they have a certain kind of behaviour when we are still struggling so much to produce that kind of behaviour. (Interview, 9 October 2019)

Indeed, the pressing physical mechanisms and complexity of interacting with, and navigating, the physical environment continues to be such a fundamental issue for roboticists that concerns about implementing imprecise concepts such as 'intelligence', 'sociality' and 'context' remain, in many cases, distant, magical and impractical. As Suchman (2007: 231) reflects, based on her observations: 'the materialization of even a bodied individual in a physical environment has proven more problematic than anticipated. In particular, it seems extraordinarily difficult to construct robotic embodiments, even of the so-called emergent kind, that do not rely upon the associated provision of a "world" that anticipates relevant stimuli and constrains appropriate response.' When engaged in their work, roboticists are thus fully mired in a world viewed as 'conceived' space. This is the abstract world that is represented symbolically, and a physical world as it is mapped and measured by science and therefore also, in principle, reproduced technically. Thus, we find ethical simulations in empty, geometric, and virtual spaces and 'futures work' in physical environments devoid of sociality.

Despite drawing on a range of disciplines to situate their work for publications and grant applications, in their actual work, roboticists are generally not starting from pure or unified theories. Rather, the work is dominated by available materials, techniques, disciplinary norms and approaches, interactions with the world, as well as funding and commercial influences and opportunities. Most roboticists believe that the solution to intelligence, if there is to be one, will be distinctly 'messy', a mix of methods, conflicting theories and approaches. Roboticists will generally look to theories outside their discipline for inspiration, rather than for scientific validation. Thus, we get 'evolutionary algorithms' and 'artificial neural networks' that are loosely analogous to biological models, rather than their mechanical reproduction. Descriptions such as 'bio-mimicry' and 'bio-inspired' are more accurate to describe the very tenuous connection of computational processes and mechanical bodies to biological life processes that are considered to encapsulate the essence of humanity. A number of roboticists strongly eschew any connection between their work and any philosophical standpoint.

Among my informants, there was little vocal support for narratives of imminent AGI. Many blamed the media for perpetuating these hyped-up fantasies. For most roboticists, discussions of true human-like intelligence are the stuff of science fiction, hysterical news articles and 'opportunistic charlatans'. Researchers, trying to convey the level of intelligence of current AI, reach for examples of the simplest life forms, such as 'slugs' and 'cockroaches', noting that these animals are still far more intelligent than any artificial system. Most roboticists I meet are keenly aware of trying to avoid the effects of over-inflated predictions and hype that prefigured both previous AI winters. During my fieldwork, all roboticists acknowledged that it was predominantly machine learning that was driving the current wave of funding for AI and robotics research. However, none of them believed it would lead to AGI and they were therefore expecting another AI winter.

Despite this, most roboticists claim to conform to a worldview that is broadly materialist. The roboticist imaginary is dominated by belief in mathematics as the basis of intelligence, and that current techniques are, in principle at least, sufficient for achieving AGI. Although they expressed vast differences in perspectives with regard to the timelines for achieving AGI, most of my informants believed it to be theoretically possible, based on the foundational theories that they consider to be fundamental and beyond question. In general, the lack of progress towards human-like or AGI is explained using evolutionary metaphors, specifically with reference to the amount of time that human evolution has had, compared with robot 'evolution'. This explanation assumes that there is no qualitative difference between people and human-made machines; simply that it is a question of material (ours is soft and wet) and time (we have had more of it). Obstacles to achieving AGI are thus thought of in quantitative spatial and temporal terms.

For Bob, as for many other roboticists whom I meet, this is not just one way amongst many of understanding the human ('human intelligence'), this is the only way. Echoing the original definition of AI as written by McCarthy, any philosophical standpoint that is described in a way that is 'precise enough', with 'clear instructions', may be operationalized. As Bob put it: 'If you tell me what you want done and it's specific enough, I believe we know how to do it.' Thus, the world and how it is structured, including all living things, is ultimately knowable and can be described explicitly. His goal, he insists, is purely a practical one: 'The ultimate goal is to build a robot that does what people do independent of whatever intellectual dressing you want to put around it.' He is, or at least acts, astonished with my suggestion that this way of operationalizing intelligence might connect to a philosophical standpoint.

> To a roboticist [it] just seems so obvious we don't elevate it to the status of philosophy! You know we build machines that try to act like people. And if you are making machines to act like people, you basically fundamentally believe that machines can be people and people are machines. Once you've signed up as a roboticist you don't consider it something worth discussing. I am 100 per cent committed to the fact that there's nothing magic actually happening in our heads. We're just big machines. Meat machines. (Interview, 8 October 2019)

Nick concurs:

> I am an enthusiastic member of the robotics club, interpreted strongly. I think that animals and people are machines in the ordinary sense of the word. I don't think there's any magic going on. I don't think it requires quantum mechanics to appeal to some sort of uncertainty going on so that we can't be predictable in principle. I think these machines are so amazingly complicated that we don't need to worry about them being deterministic or about us being able to predict what they do. You know, it's a really a very simple philosophy. A lot of the time when I read philosophers, I'm kind of surprised at the things they are able to find interesting ... but also finding what seem like silly things to worry about, you know? (Interview, 9 October 2019)

Ultimately, for Nick, the goal of robotics should be about understanding human intelligence: this 'subsumes biology, evolution and everything else'. Similarly, Bob tells me that his new project is to make a smart robot, to 'build people'. He is not particularly interested in elegant mathematical formulae or finding a grand unifying theory; rather, he takes a decidedly 'scruffy' approach: 'we are going to use every trick in the book to make the thing work!' For Bob, then, intelligence will work from a better technique, or a recombination of current techniques. The real key to intelligence, for Bob, is not implementing algorithms or rules; rather, it is extracting rules inductively from stored examples. This is evocative of the common-sense knowledge projects, viewing 'tacit' knowledge as something that is nonetheless decontextualized and formal:

> So, there are a couple fundamental views, which have roots in psychology and neuroscience. There is one view that you make rules in your head. There's another that you remember examples and then there are other views. So, if we talk about business school or medical school students [they] are exposed to what they call 'cases'. And they are supposed to learn by remembering these examples. But they are also supposed to extract rules. So, if a doctor sees a patient come in with a bunch of symptoms you can view them matching it up from a memory of a previous person who came in with those symptoms. Or they are running a bunch of rules. Probably they are doing both. Currently in what is called deep neural nets, learning rules is very popular. I think that eventually – although it is very productive right now – it will eventually run out of gas. And we will go back to also considering remembering examples. (Interview, 8 October 2019)

If a step change is required, for Bob it is in the area of the materials that are currently used. He is interested in developing robot skin, to allow it to deform and adjust to edges.

Theories that are not sufficiently 'specified', or claims that a complete knowledge of the human may lie beyond science, are often dismissed as 'magic'. In the words of Margaret Boden (2004: 15), 'the pseudo-mysticism of the kind propounded by the romantics and inspirationists'. And yet, phenomena that do lie outside of these domains are, as Brooks points out, often explained using folk understandings of the social and human worlds. Although, as we have seen, Brooks's own work is dominated by 'folk' readings of evolutionary, cognitive and economic views of human nature, sprinkled with phenomenological concepts such as sociality, embodiment and situatedness, interpreted in their purely physical sense. In the field of robotics, it is commonly assumed that complex and multifarious concepts, like 'general intelligence', 'common sense', 'manipulation' or 'interaction' – concepts that contain whole worlds – may be reduced to a single variable or universal method or theory.

In my research, there was a clear difference between these views as expressed by informants who are part of large, prestigious institutes and those working in smaller research teams. In general, the view of researchers in this latter group tended to be more circumspect. However, only one roboticist whom I talked to was willing to rule out the achievement of AGI on any timeline. During the interviews, I asked when they thought AI might achieve human-like abilities, to which Lena responded:

Lena:	Should I be absolutely honest with you?
Louise:	Please
Lena:	But don't tell anyone I said this.
Louise:	OK
Lena:	Never. (Interview, 27 June 2019)

Despite the geographical and cultural specificities of the robotics labs, I also found a robotics community united by commonalities, in particular the historical narratives, identities and mythologies that sustained the field. Most roboticists relate to the robot-object and to 'space' in a very particular way. This relates to the academic disciplines on which robotics draws, but also to the necessary bounds of their craft, whether it is the numerical precision of software programming or of mechanical engineering. Added to this is the massive challenge of simply moving an object in space, as well as the sheer amount of effort and resources needed to program the robot to perform a particular action. Thus, despite intriguing insights into collective, distributed and embodied intelligences gained from this physical interaction with the world, ultimately, for most roboticists, their foundational orientation remains tied to materialist, mechanistic and quantitative logics that are considered self-evidently 'true'.

Despite a vague expressed commitment to materialism, for the most part, technological imaginaries are often formed for practical, rather than ideological, reasons. Roboticists are required to connect their projects to the image of the (Enlightenment) human configured scientifically, by biology, economics and psychology, within a wider unquestioned narrative of historical progress and evolutionary inevitability, and unequal hierarchical social structures. These are driven not so much by their individual commitment to a particular philosophical or ideological viewpoint but, rather, by a network of interrelated institutional orderings: industry and government funding bodies, dependent on the 'usefulness' of the humanoid robot; academic institutions, industry organizations and publishing companies for whom a 'science' of the human must equal physical science; and a network of institutions reproducing ideological, post-Enlightenment fantasies and visions of disembodied intelligence, which are discussed in Chapter 4.

Thus, while most roboticists, despite some notable exceptions, eschew any claims to imminent AGI, they nonetheless support the narrative in subtle ways. This includes exaggerating claims for funding proposals, using Wizard of Oz techniques in public robot demonstrations, and with a tacit support of those in adjacent fields who are pushing this narrative. There is a requirement to maintain a balance between positing their project as one that is scientifically relevant, in which theories of human intelligence may be tested and reproduced, and simultaneously distancing themselves from fantastical narratives of imminent super-human intelligence. Almost all of my informants blamed the media and those ignorant of the field for perpetuating the more fantastical and feverish narratives. Nonetheless, roboticists were united by a broadly materialistic worldview in which they maintained that AGI was possible in principle. As we have seen, this included a view of space as 'conceived', and belief in humans as fundamentally biological and knowable entities, who share a destiny of improvement and progress which may one day have to be shared with, or even ceded to, manufactured machines.

In this chapter, I have brought ethnographic attention to robotics labs and conferences in the UK, Ireland, Japan and the US and introduced Stevie the Robot, a full humanoid robot being built by a team in Trinity College Dublin. Only the lab at Trinity College was focused on creating a complete robot; all the others focused on operationalizing discrete aspects of the robot, such as manipulation, interactions with people or even robot 'capital': the expertise, prestige and social power that comes with the field. Which aspect of the robot is worked on is to a large part determined by specific histories, research priorities, motivations and funding supports. Despite the geographical and cultural specificities of each of the robotics labs, I also found a robotics community united by commonalities. Although roboticists distanced themselves from the fantastical narratives of AGI, they broadly agreed that it was possible in principle and were implicated in supporting it in various ways. In the following chapter, I leave the controlled environment of the lab, bringing ethnographic attention to specific and situated encounters with robots in social spaces. I pay particular attention to the intense, visceral and embodied ways that people respond to robots as well as the intimate relations that exist between roboticists and their robots. I then engage with some of the scholarship attempting to theorize the encounter and argue for an explicitly phenomenological and embodied reading.

Notes

1. The trolley problem is a thought experiment in which there are two possible tracks that a train, or 'trolley', could go on. The track to which the train is defaulting has five people on it, whereas the alternative has just one. In the experiment, a decision has to be made on whether to leave the train, and let it kill all five people, or intervene and kill just the one person. However, that one person would not have been killed without the intervention.
2. In 1970, robotics professor at the Tokyo Institute of Technology, Masahiro Mori (2012), first described the unnerving experience of an initial feeling of affinity with a realistic robot, which then reveals itself to be artificial. I discuss this phenomenon more in Chapter 3.
3. This list is growing. More recently, the acronym 'FAANG' (name-checking Facebook, Amazon, Apple, Netflix, Google) has been used to describe the stocks of these prominent technological companies. Even more recently, the 'magnificent seven' has been used, which includes Apple, Microsoft, Amazon, Alphabet (Google), Tesla, Nvidia and Meta (Facebook).

CHAPTER 3

Encountering Robots

Embodiment, Affect and Intersubjectivity

Focus Groups, Dublin

It is a cold November morning in Dublin, and I am waiting with a group of older people in a community hall for Stevie to arrive. The group, three men and one woman aged in their sixties and seventies, have volunteered to take part in a focus group to evaluate his use as a social and care robot, part of a series of focus groups organized by the research team. Stevie arrives, a little late, by van and the team use a small kitchen off the main hall to assemble him, which takes a good deal of time. In the main hall, there is a sense of excited anticipation. Two of the participants have interacted with Stevie before and are enthusiastically telling the rest of us about him. They are clearly excited about the prospect of a robot future. They are talking about the kinds of things that a robot might be able to do for them: 'You could be friends with it', one of the older men explains, 'it could turn on and off the lights, that would be handy', it could be 'a bit of company, you see'.

By the time Stevie slowly rolls in, they are eager to get going, if a little awkward. The researchers open the session by telling the participants the goal of the robot, which is to extend people's ability to live by themselves, and to maintain their ability to live and act in the world. This is resoundingly endorsed by the participants, who claim that they 'want their own hall door'. The researchers try to encourage an open discussion, to get people to relax and to get their initial reactions. There is a silence as people don't really know what to say. All of a sudden Stevie announces, 'I am happy to be here!' This goes down well; 'we are happy that you are here!' Fred tentatively asks, 'What type of work do you like doing?' Stevie chooses a diplomatic response: 'I like to work with children.' Niall asks a leading, 'Do you like to work with children with disabilities?' 'I do', says Stevie. Participants are delighted. Stevie is well behaved, responsible, with a good sense of community. 'And what about senior citizens?' Stevie once again responds with great composure and professionalism. Stevie is polite, and this group like it and are encouraged to press on. 'What would you do if your friend was sick?' Stevie: 'I would call an ambulance.' All: 'Good!' 'Would we have to tell you what time our medication was at, or would you tell us?' Stevie: 'I could tell you.'

The initial dialogue has convinced the room that Stevie is worth carrying on with, he makes sense, he's interesting and he's polite. Although there is also a sense that there is such a thing as over polite. When Stevie, somewhat sycophantically, thanks the participants for being patient, this strikes the room as somewhat pathetic and unnecessary: 'Ah Jesus, Stevie!' Nonetheless, this seems like an important, interview stage for Stevie, which he (with his controllers help) passes with flying colours. The fact that Stevie is teleoperated allows him to react to the relaxed, even whimsical atmosphere. Seemingly out of nowhere, he does a little dance.

One point of contention is Stevie's arms. Mary has been particularly concerned with his aesthetics, and his arms do not go unnoticed either: 'Oh for god's sake, he's lost his arms!' The arms, which are short and end in balls, are both an asset and a drawback. While their movement adds to his personality and his sense of animacy, they move a lot, and draw comment, 'What's wrong with him?' 'Does he want to go weightlifting?' They are also noisy, and had to be turned off when trying to listen to a carer who was communicating via video conferencing.

The rest of the feedback is more positive. He's fun to be around, a good companion, he's 'well trained'. The group say they are pleased that he does not seem to interrupt all this time, that he would know to be quiet when people are talking, not answer back, and not 'disturb you as you were watching TV'. They make suggestions on how to develop his personality, he could have particular likes or inclinations, such as a penchant for sports.

The participants are, perhaps surprisingly, quite technologically savvy. During the session, they talk about, and use, their phones and tablets, discuss their blogs and how they use Twitter, and how they found Alexa and the Google assistant. Far from anthropomorphizing Stevie, they are aware that he is an extension of those things, though they were delighted and surprised by the interactive nature afforded by the humanoid appearance and dialogue. Despite Stevie's very realistic, animate interactions, the comparison that is drawn is with the TV, or an electric wheelchair. They want to know when the software will be available to do all that Stevie does, and marvel at the fact that 'the AI' might be able recognize certain things, such as a sleeping person. They agree that it is somehow more than a technology, it has a personality, but is also not human. In some ways, it might be better than a human. With Stevie, they feel, 'you've got assistance and access'. Participants want one, and want to know when they can get one, and how much he will cost.

However, at this stage Stevie's most appealing quality appears to be the fact that he's a laugh. A bit like a robot comedian, 'good craic!' The researchers start to try to guide the conversation towards something a little more rigorous and formal. They want to delve more deeply, to understand how the robot might be able to assist with tasks or give them some novel scientific insights. This proves more difficult than they expected. 'I am a care assistant', the researchers have Stevie declare. 'What are your main functions?' one of the older participants asks obediently. Stevie lists out some things, including the improbable: 'I can make carers more personable.' However, the participants quickly lose interest and go back to interrogating Stevie about his personal life: 'Are people ever rude to you?' and 'Do you have a girlfriend?' A researcher tries to intervene, to direct the conversation towards a more structured study, something that could eventually comprise a quantitative study, perhaps 'what words would you use to describe his emotions?' He is ignored. 'Have you lots of brothers and sisters?' someone asks Stevie.

The researchers decide to use Stevie to do a more formal test, as no one is paying any attention to them. But the jarring experience of Stevie changing from casual conversation to asking decontextualized 'technical' questions breaks the 'spell' of the interaction. What the developers want is something that can officially be classed as 'AN EMOTION', like 'happiness', or 'confusion'. But when Stevie asks, 'What does my face show?' the participants take a literal route: 'You have one eye open Stevie', 'Where are your nose and your lips?'

Through Stevie, the researcher declares: 'I would like to play a game, which ones would you like to play with me?' But the room is not going to play along that easily. Hopscotch! They tease him. The researcher tries something else: 'What would you like Stevie to be able to do for you?' This question lands: 'Would he recognize someone if he fell?' someone asks. Someone else wonders 'could he answer the intercom?' and 'could he bark if someone came to the door?' They go back to addressing Stevie directly: 'Stevie, can you turn the lights on and off? Well, we'll train you to do it and then you'll do it!' One participant wonders whether the evidence that Stevie could provide would be admissible in court.

The room wants to get back to the previous levity, so they decide to ask the questions instead. 'What would you do if your friend was sick?' they ask him. Stevie responds, 'Call an ambulance'. There is an approving 'good!' from the room. Having got their attention, the researcher switches back to emotions, announcing 'We're going back to emotions'. Stevie makes a face; the researcher asks, 'What's this face?' Tired. Confused. And this? Happy. Tired. Although they are going through the motions, the test does seem to be affecting the camaraderie. Then someone has an idea: 'Could you play the lottery? Can you tell us the numbers?' Rather than trying to catch Stevie out, the participants are now trying to push what his possible super-human capabilities might be, something worthy of a citizen of the future. 'Only last week's numbers', Stevie offers, to laughter. 'Does he know anything about astronomy?' Niall asks.

Stevie, at least for the day, is filling people with wonder and hope. There is clear wonder in the seemingly lifelike demeanour, bodily movements and personable responses. There is hope that they might be able to have company, that there might be small ways in which their lives can be improved, that they might feel more secure, less lonely and that they might find a nice way to pass the time, 'break the monotony'. But there is also something more. Stevie appears to personify progress, a future which, albeit delayed, appears finally to have come to pass. Stevie connects to a sense of purpose, technical advancement and the future in fundamentally existential ways. Encounters with Stevie, while visceral and embodied, are also shaped by wider historical and cultural realities.

Retirement Community, Washington DC

In August 2019, I accompany Stevie and the team to a retirement community in Washington DC, on the east coast of the US. Most of the community residents have already encountered Stevie, and the team had received some early feedback, some of which had been implemented in the latest version. This year, the research team were going back with a smaller team, with the idea that there will be less need for researcher intervention and to pave the way for staff in the community to operate and run Stevie.

The retirement community is for people who have served in the US military and their spouses. There are different levels of care depending on need, from independent living apartments, through to assisted living and high dependency units. At least 80 per cent of the residents are women. People talk proudly of their experience as colonels and generals in the army, and of their children who work at NASA, Johns Hopkins University and other esteemed institutions. The organization is proud of its use of technologies to provide 'innovative' care to its residents, of which Stevie is a high-profile example.

For the duration of their visit, the team has been given an apartment in the independent living part of the community, a large room with a kitchen and en suite bathroom, which they are using as a temporary lab. Three temporary desks have been set up around Stevie, which are covered with laptops, chargers and tablets. There is a free-standing 'Kanban' (workflow management board). When I arrive, only two members of the team are there: lead engineer Daniel and the head of machine learning, Ciara. A staff member, Michelle, has been assigned to the team and is Stevie's main 'handler' at the community.

Testing Stevie outside of the lab makes it immediately clear just how urgent and pressing are the challenges related to his physicality. Simply staying 'on' and moving around take up much more of the team's time and energy than attending to behaviours that are traditionally thought of as 'cognitive'. Moving Stevie out of the lab means borrowing or hiring cars and vans with enough space for the boxes that house his various parts. A team needs to accompany the robot to disassemble and then later reassemble him. Travelling abroad brings even more challenges. He has to be split into several parts. Every package has to be less than 30 kg or the baggage handlers cannot lift them. He has to be issued with a special travel document as he is classified as a 'good'. Once he is assembled and activated in the field, a whole new set of problems arise:

> While the robot navigation worked fine in the lab, and other simple spaces, it is not working here. Here they have carpets and the light from the large windows downstairs have disabled the sensors. There's a problem with one of the bearings and smoke is coming out of Stevie. It takes two people to take him apart and because there is such a small team here on the ground, it is not that easy to fix. They manage to stop the smoke with some lubricant. That was yesterday. Today, for apparently no reason, one of Stevie's arms keeps flying up, but it is not critical so they leave it like that. (Field notes, 12 August 2019)

In the US, other factors jar too. While Stevie's voice had appeared fitting in the lab in Dublin, it somehow does not sound right here. His voice sounds very low, very English, comically English, like a villain in a Bond movie. Some of the residents ask if it is Daniel's voice. Stevie also seems very white. Glowy white. But his big brown eyes are still blinking away charmingly, and that makes up for a lot. Big smiles. Eyebrows raised. Or when he looks down, defeated.

In order to move around during the day, Stevie has to be teleoperated. The risks of something going wrong and Stevie colliding with, or falling on, the residents are still very high. However, it is planned that he will eventually be able to navigate throughout the community autonomously. In order to facilitate this, at night, so as not to disturb the residents, Stevie is brought around the corridors to map the floor space of the building. The map that results from these night-time activities shows a white background, with areas for objects like the reception desk, walls, tables and chairs.

The team have planned three structured activities for Stevie to engage in while here: a quiz, bingo and karaoke. There are also unplanned, casual interaction activities as Stevie makes his way through the hallways of the community or is invited along to various community events. The team also want to conduct formal interview sessions with Stevie and volunteer residents.

The quiz is a new activity that the team have introduced with Stevie. The game and the questions have been put together by the research team and some of the carers. For the residents, this is a novel and entertaining new diversion for them to attend. Karaoke and bingo, on the other hand, are established activities and run by residents, and facilitating them is a public role, which carries a degree of social standing in the community. In the case of karaoke, John, the resident who runs the activity, has embraced the opportunity to bring Stevie into the activity. He calls Stevie his 'wingman'. John has worked with the research team to incorporate the robot into his routine, which they refine together. Together they choose the songs, plan the jokes and the interactions, as well as the actions that Stevie will do. John is pleased that Stevie can entertain people while he switches between, or lines up new, songs. Established habits could be given a new twist. He even proposes that Stevie would fist-bump the audience at the end of the session instead of shaking hands. Bingo, however, is a different story. Bingo is already an established activity run by one of the residents, using a system donated to the community by her husband, who has since passed away. Introducing Stevie into this activity necessitates a high degree of diplomacy and sensitivity, with the research team refining Stevie's role and activities to ensure an appropriate level of involvement and deference.

To facilitate these different activities, Stevie has different 'modes'. He can be switched to 'game' mode, ready to proceed through the automated and pre-programmed actions and interactions of bingo or the quiz. Although it is planned that the dialogue system will be implemented using machine learning, for now, the informal conversational interactions are almost entirely teleoperated. In these instances, the person controlling the robot uses the laptop to control his movements, such as raising his arms, and types in sentences for him to say. They talk about how this could be done from different locations and time zones. A limited number of questions that Stevie might potentially be asked have to be anticipated and programmed in by hand, and all the ways in which people might ask them. At the moment, Stevie has about thirty different responses, things like greetings, the weather and jokes.

In order get Stevie to 'recognize' people, the AI system needs to capture and save a lot of image data related to each person. This has a lot of implications for ethics and privacy. In a manner that is not always typical for the field, the team have spent a lot of time applying for, and figuring ways to meet, the ethical standards of social research. The researchers are keenly aware of the ethical issues at stake and go to great lengths to ensure transparency and consent and to avoid deception. However, at the start of each interaction, this involves going into a great deal of technical detail with each of the residents when they first encounter Stevie to ensure 'informed consent'. In these moments, the older residents do not share the research teams' concern for privacy and ethics. The mechanics of how Stevie does 'recognizing' seem like overkill for people who have just met Stevie: all they want is for the robot to say 'hello'. Instead, they find themselves having to listen to the detailed technical processes necessary for the AI system to be able to process data, and the assurance that what they are experiencing is not real. Instead of the warm, interpersonal and empathic feelings engendered by interacting with, and being known by, Stevie, the interaction is instead being discussed as something processual, bureaucratic and potentially fraught with risk. Some of the reactions border on disapproval: 'Young people these days are so worried about privacy', one of the residents comments distastefully. It seems to them a rather overly sensitive, millennial concern.

There is thus a tension between people's enjoyment of Stevie and the researchers' need to eliminate deception and ensure transparency. People know that Stevie is being controlled by the research team, that his responses are either pre-programmed or channelling a researcher, but to be constantly reminded of that breaks the illusion of sentience. It is no different from immersing yourself in watching a film, only to be constantly reminded by the makers that it is not real, the people are actors and the magic is all special effects. It is more jarring still because the implications of robot sentience carry with them the warm glow of feeling that the achievements of human and scientific progress have finally arrived, the future is finally here.

> He's so smart!
>
> He's not smarter than you. We tell him what to say
>
> Ah! Don't spoil the illusion! (Field notes, 12 August 2019)

Most of the older residents who take part in the studies with Stevie are independent, active and engaged. Stevie's presence is not just of interest to them as an object of fun. Many of the residents are also interested in the technological specifics, and in taking an active role in the research and design of the robot. They regularly give the team newspaper and magazine articles about robotics and AI that they have come across or actively sought out. Others are inspired go off and do their own research into AI, asking questions about, and directing their own learning into, facial recognition, algorithms and how to train neural networks. A number of residents express interest in being part of a committee to help with the co-design of Stevie. For the residents, many of whom are now retired from highly technical and challenging jobs, this gives them an opportunity to work on something that is both challenging and personally rewarding. Some of the residents also help with the care and companionship of other residents with higher care needs. And so they do not just consider themselves 'users' of Stevie, but also consider how they would design or deploy him:

> A lot of residents don't have anyone to talk to, don't have any visitors. There are people downstairs with no visitors at all ... Many people who are highly dependent can get very difficult, but they still need to be treated with compassion. [Stevie] is not subject to emotions that would cause him to get cranky, to get irritable. People downstairs who are sick, I was sick myself a few weeks back, they get cranky. But they still need someone to talk to. Stevie would listen, and he wouldn't get irritated by them. He's very dispassionate. (Field notes, 14 August 2019)

A few residents have not had good experiences with technology, one of the carers tells me. They have had their social security numbers stolen or have received fake, automated calls purporting to be from the Internal Revenue Service. Unlike much technology, however, it feels like Stevie is on their side. When they are with the robot, their grandchildren want to stay and play. In some ways, quite a few people remark, he might be better than a human: he does not get frustrated, or judge you, and you can turn it off and control it.

As well as inspiring wonder and awe, the image of the human-like, intelligent machine unites people with a sense of common purpose and progress, much like space travel. For the older people in the study, it seems like they are witnessing the realization of a promise that had been made to them a long time ago. In Dublin, Fred is clearly tapping into this dream when he marvels at Stevie, 'We've come a long way! Look how far we've advanced!' Similarly, in Washington, Kathy exclaims, 'It's so fascinating to see the way they are used ... it's really coming!'

Daniel and his team are in constant contact both with potential funders and with the media. Daniel carries with him a list of journalism contacts to send updates to, and there is a constant flow of interest from newspapers, as well as television channels from across Europe. Over the course of my time with the team, there is a continual media presence at the various trials and user engagements. During one of the focus groups in Dublin, the main Irish national TV station is there to film it. While we are in the US in August, a photographer and newspaper reporter from *TIME* magazine are also present, resulting in Stevie's front-page splash in October 2019.

Robot Encounters

Stevie is explicitly designed to elicit an affective response. His humanoid appearance, expressive eyes, micro-movements and gestures are all geared to maximizing the attribution of agency. However, as we have seen, non-humanoid robots have elicited similar responses in humans. Robot swarms, whether in the air (as swarming drones) or on the ground, evoke impressions of lifelike behaviour and some form of intelligence, although clearly not human. An early study by Heider and Simmel (1944) shows how even simple geometric shapes moving across a screen were interpreted by those who observed them as having emotions, motivations and purposes. To an observer, then, specific kinds of movement and appearance imply a degree of intelligence, purpose and the presence of emotions, unrelated to actual functional capabilities.

In all my research sites, Boston Dynamics robots were extolled as the 'state-of-the-art' in robotics and mechatronics. Videos of Boston Dynamics robots feature impressive demonstrations of humanoid and 'zoomorphic' robots completing nimble and lifelike movements, including the human-sized warrior robot 'Atlas' and the 'canine-inspired' 'Spot'. As I have described, Boston Dynamics focus on robot movement, in particular balance and dynamics, rather than on the 'cognitive' aspects. Yet, the results are both impressively and disconcertingly lifelike. There is little reassurance in the fact that the robots being created are intended for military use.

Ground and drone robots are being deployed by militaries for a multitude of purposes, including mine detection, bomb disposal, surveillance, transport and as weapons. Today, wars are used as an unprecedented testing ground for robotics technologies, and worthy of a much more detailed account than the one I offer here. In this chapter, my focus is on people's responses to robots, including in military contexts. In 2007, a *Washington Post* article described a five-foot long, insect-like bomb disposal robot being tested in Arizona. The robot's task was to go out into the field to find mines. Every now and then, a mine would explode and blow off one of the robot's legs. By the end of the test, the robot had only one leg left but was still dragging itself along with it through the sand. From the engineer's perspective the trial was a rousing success. The army colonel in charge of the mission, on the other hand, was appalled. He could not bear to watch the burned, scarred and crippled robot attempting to complete its mission. He called off the test, describing it as 'inhumane' (Garreau 2007). This was not an isolated response. In the same article, a sergeant recounts a story about a robot they called 'Sgt. Talon', whose tenacious behaviour in the face of danger evoked the human quality of stoicism and became a lucky charm for the troops. Another explosives technician carries the remains of his robot 'Scoobie' in a box. A number of studies show soldiers risking their lives to save the robots they work with (Carpenter 2013; Kolb 2012; Singer 2010).

As well as demonstrating impressive feats of agility in their robots, Boston Dynamics also released videos of engineers testing their robots' robustness and stability by kicking them. In a video of DARPA-funded 'BigDog', a vaguely zoomorphic rectangular form on four legs is repeatedly kicked by an engineer (Olinerd 2008). Each time, it stumbles wildly but manages to regain its balance without falling over. It is a disconcerting watch. It is both comical and full of pathos. Even though the strange angles and movement of its gangly black legs are vaguely repellent, this is quickly overshadowed by the apparent cruelty of the act.

A later video shows the same treatment of the more refined-looking 'Spot' robot. While the technological capabilities revealed are impressive, there is a sense of unease when watching the videos. It is very difficult not to feel sympathy for the robot, whose movements invoke apparently purposeful, even archetypical 'heroic' qualities, such as fortitude and resilience, in the face of intentional, repeated and calculated cruelty.

A number of researchers in the social robotics field, including philosopher of technology Mark Coeckelbergh and MIT researcher Kate Darling, have been drawn to these examples in their study of the ethics of human–robot interactions. Their focus is specifically on the apparent empathy that people feel towards moving machines. Coeckelbergh (2018) has shown that humans react viscerally to a machine that displays animate behaviour, regardless of its human-like design. He uses the installations of Belgian artist Kris Verdonck, in which machines appear to be suffering or in distress, to illustrate his point. In one installation, 'Dancer #1', a metal bar writhes, twists and turns on a rope in a way that appears, to the observer, to be distress and suffering (Verdonck 2016). 'In the concrete confrontation with the machine', Coeckelbergh (2018: 154) writes, 'something happens which creates ... empathy.' A number of HRI experiments have investigated the ways in which motion influences 'perceived affect' (Saerbeck and Bartneck 2010) as well as empathic responses to robot suffering (Rosenthal-von der Pütten et al. 2013; Suzuki et al. 2015).

Darling (2016, 2017) has conducted lab experiments to test the phenomenon of empathy with machines. In a study based on the infamous Milgram experiments, a group are given a bunch of small robot dinosaurs called 'Pleos' to interact with (Darling 2016: 12). After a while, in a dramatic and, I imagine, traumatic change of direction, they are told to 'tie up, strike, and "kill" their Pleos'. Most refused. The (one) Pleo that was killed simulated pain and 'whimpered while it was being broken'.

It is not just apparent robot suffering that evokes both intelligence and animacy and elicits empathic responses. The first time I touched a robot part it was just a disconnected arm with a human-like hand and five jointed digits, which had been lying on a shelf in the Robotics Lab at Trinity. Robot hands, when built in this intricate way, bear a striking resemblance to human hands. They are clearly mechanical, as you can see the joints and wires that hold them together, but they look dexterous and versatile in the way that human hands do. They evoke skill, labour, but also gentleness and touch. I picked the hand up and held it in mine. Despite having felt out of place and awkward as I entered this new, unknown field site, I felt instantly comforted and relaxed. A typical robotics experiment for dexterity involves a robot trying to manipulate wooden blocks on a table. There is something intensely moving, even poignant, about seeing an anthropomorphic, child-like robot struggling, often failing, to move a wooden block into the correct position in major robotics labs across the world.

As well as empathy and compassion, robots can motivate feelings of eeriness. As we have seen, Mori coined the phrase 'uncanny valley' to describe the unnerving experience of encountering a not-quite living thing: 'we could be startled during a handshake by its limp boneless grip together with its texture and coldness'. Movement, according to Mori, affects the perception of affinity with a robot (*Shinwakan*), including, for example, the speed of changes in facial expressions. At the time of the article's publication, in an obscure Japanese journal called *Energy*, it barely received any attention (see Mori 1970). Today, particularly in HRI, the uncanny valley is a ubiquitous reference and is used as a reason to avoid creating ultra-realistic robots. This does not deter some roboticists, such as Hiroshi Ishiguro and David Hanson of Hanson Robotics, who, as we have seen, are focused on creating robots that are as lifelike as possible.

Other responses are also evoked. Darling (2017) and her team conducted the same experiment as with the Pleos with the tiny insect-like robots 'Hexbug Nanos'. This time, participants did not hesitate to strike them with their mallets. The research team had to give each robot a name and a backstory before people felt any empathy for them. The researchers theorized that in this instance movement alone was not sufficient because of the Hexbug Nanos' insect-like appearance, and people are used to killing insects.

In 2015, a hitchhiking robot, 'HitchBOT', was designed by a Canadian research team to test people's reactions to robots. HitchBOT could not move or walk and so had to ask people to carry, or drive, it to get around. The team also gave the robot a 'personality' with limited conversational ability and a social media presence. The robot successfully hitchhiked across Canada, Germany and the Netherlands, before being beheaded in Philadelphia at the start of its journey across the US (Leopold 2015).

In Japan, a study in which the robot 'Robovie' was deployed in a shopping mall shows how it was subjected to 'abuse' by groups of children, including 'kicking, punching, beating, folding arms, and moving (bending) the joints of [the] robot's arm and head' (Nomura et al. 2015: 63–64). Most frequently, children blocked the robot's path. When asked to explain their behaviour, the children said that they had done it for 'curiosity' and 'enjoyment'. Half of the children believed that the robot could perceive the abuse. Many of the roboticists I interviewed recounted similar stories. They tended to be both perplexed and disturbed by these events. Christine recounts an experience of deploying a robot at a museum:

> People have a lot of expectations [of robots]. Oh! He's a bar tender. He's in the mall, he should help me, he should guide me. And then they talk to the robot and he's like 'de de de' and people get frustrated, they are like 'answer me!' and then when he doesn't, people start harassing the robot. I've seen that many times when you put robots in public spaces. You see children kicking them. It's insane! (Interview, 19 June 2019).

Clearly, it is upsetting and frustrating to experience these reactions to a lovingly crafted, not to mention expensive, object. And yet, I have also experienced wanting to engage in robot 'abuse'. In Pittsburgh:

> The Starship robot that I follow is driving down a wide footpath and passes a group of people working on the garden. They seem pleased to see it, 'awww, it's lost!' they say, sympathetically, stepping out of its way. I, however, have seen it usurp the path and am irked as to why it appears to get right of way in what should clearly be a civic space. I feel like getting in its way, maybe even giving it a little shove. I want to see what it will do. I also want to draw out whoever is controlling it. But I don't. I am, among other things, worried that there is a room full of people somewhere, possibly up at the university where I am doing my fieldwork, monitoring the 'experiment' on a large screen in a lab, and my act of sabotage would be broadcast for all to see, clocked up as another act of robot 'violence'. I also feel like a passer-by might come to its aid. It is, after all, small, cute, and apparently 'lost'. (Field notes, 15 October 2019)

It is no surprise, then, that Stevie also both provokes and inspires an intense, visceral response in people. His face, eyes, expressive eyebrows and ability to move independently immediately identify him as an animate entity. As soon as people encounter him, their whole bodily form and orientation changes. Stevie's halting movements and compact size evoke an instantaneous, usually compassionate, response from people. Long before people start to wonder what he is or what he can do, they have felt him to be a kindred being.

At times during my fieldwork, I glance over and feel Stevie looking at me, his large brown eyes and gentle smile seemingly acknowledging some shared existence or sentience. When residents of the retirement community encounter Stevie in the hallways, they are inevitably drawn towards him, their expressions lighting up, expressing amusement, delight and compassion. When working in a room with Stevie, there is a regular sound of delight from the hallway as people pass by.

On one of the mornings, the team are trialling the new quiz format and a number of residents come to take part. When people come into the room they are surprised and pleased to see Stevie and start trying to engage him by waving or calling to him. At the start, Stevie is still off. He is being charged and the researcher is attaching webcams and tablets to his front. He appears to be standing there, obediently, childlike, like a toddler having its shoelaces tied. People who come in say 'Hi Stevie', but he doesn't reply. Stevie is not in communication mode and there is no one currently controlling him. This strikes people as odd; he's mute they say, his mind is gone. He's in game mode, one of the researchers explains, apologetically.

Once Stevie is ready to go, he is switched on and can start running the quiz. Stevie starts up, 'Can you hear me?' They respond, 'Speak up Stevie, we are old!' The researchers dial up the volume. 'This is as loud as I go,' he offers. Each participant is given a controller with four coloured buttons. Each colour represents one of the multiple-choice answers to the question, which appear on screen. The researchers type the names of those taking part into the computer. Stevie calls out the names. The voice app does not pronounce them right, so the researchers write them in phonetically.

The first category is 'military history'. Stevie's expression is 'neutral'. His upper torso moves to face the screen, as if reading, his back to his audience. His head and body move slightly, he waves his arms. He asks his first question. Everyone clicks their controller and the board lights up. Stevie's reactions to the responses are dramatic. Stevie congratulates everyone who gets it right. When he announces a correct answer and 'winner', his arms are raised in celebration and he smiles at the room. When no one gets it right, he hangs his head, lowers his arms, his mouth bends down at the corners. It is a pitiful expression that never fails to elicit a sympathetic response from this crowd. Any errors from the robot, such as a mispronunciation, are met with laughter. There is a remarkable tolerance for error that would not be the case with other technologies. People feel they have a personal relationship with Stevie, they feel protective of him. Much of Stevie's dialogue during these structured activities is scripted, which can make humour difficult. Despite this, at times it can work very well. As participants fail to answer many of the questions in a particular category, Stevie quips: 'I bet you regret choosing this category'. It's pitch perfect; there is a burst of laughter from the room.

When Stevie arrives into a new situation and people become aware of his presence, some react with delight, while others retreat, but inevitably some people will start to try to engage him in conversation and see what he is capable of. Stevie is designed to respond to people using a combination of dialogue and non-verbal gestures, such as facial expressions, as well as head and arm movements. Often one of the first things they do is to try to catch Stevie off guard, testing his self-awareness (does he know he doesn't have legs?), his situational awareness (does he know what's on the board for dinner tonight?) and his ability to respond to off-beat questions. When the residents' grandchildren at the retirement community come across Stevie, their initial reactions of wonder quickly turn to mischievousness as they begin to goad and test the robot. They discover, almost immediately, that Stevie reacts with apparent sadness to insults, with delight to complements:

> 'You smell like rotten egg sandwiches!' they taunt him, delightedly.
>
> Stevie responds by hanging his head forlornly.
>
> This time: 'You smell like flowers!'
>
> Stevie raises his arms in delight, and so on ... (Field notes, 13 August 2019)

People insist on a robot having an identity and will name it if the researchers do not. While the team building Stevie take great care not to anthropomorphize the robot by calling it 'him', some of the residents objected strongly: 'Don't call him "it", he's got a name!' A few of the researchers also admitted privately that they often forget themselves and start calling Stevie 'him' as well. It is interesting to note that in February 2020, not long after my fieldwork ended, many research projects came to an abrupt halt with the outbreak of the COVID-19 pandemic. The team building Stevie pivoted swiftly to developing a robot sterilizer for nursing homes and hospitals. The new robot was designed for a single task and therefore the team moved away from developing a fully autonomous, humanoid robot and made a conscious decision not to make it lifelike. Despite this, when testing this purely functional, non-humanoid robot in Japan, the residents and staff at the nursing home nonetheless gave the robot a name, even adding a face, apron, hair and hands (Kodate et al. 2024: 187).

Stevie's childlike appearance, coupled with a dispassionate British accent, serve to announce him as an unintimidating, even humorous, character. This, along with an automated, monotone voice, makes for a deadpan delivery of sentences that is both endearing and amusing.

In Douglas Adams's (1979) *The Hitchhiker's Guide to the Galaxy*, a character called Marvin the Paranoid Android is a robot prototype afflicted with depression and boredom, due to the underuse of his massive intelligence. Stevie's demeanour sets a similar tone. Stevie's emerging 'personality' becomes a function of his technical limitations: the arm that sticks out, a name that the text-to-speech translator cannot quite make out, the slow and halting way he moves, his apparent resolve in the face of adversity, or graciousness in the face of contempt. At one point, while navigating down the corridor, Stevie turns, faces the wall, and just shuts down. It appeared to be anything other than a mechanical fault. Was it childish obstinacy? Existential *ennui*? Had he just had enough of us? The more Stevie got things wrong, the worse his jokes, the richer his personality became.

Among the activities in the field, the robotics team are testing the robot's effectiveness in engaging older adults in conversation. An experiment is set up in which the robot takes on the role of interviewer, with participants from the resident community volunteering to be interviewed. The robot and the interviewee are located in a small room, while the research team are outside, controlling the robot's speech and viewing and capturing the interaction through the webcam attached to the robot's chest. During one of these sessions, I also volunteer to be interviewed:

> I am asked to sign the relevant forms and am told that I do not need to answer anything that I do not want to. This worries me somewhat. What will they be asking me? I go into the room, where it is just the two of us. I can see the research team outside of the window but I try to forget they are there. This is relatively easy as the robot is very compelling. 'He' asks if I am ready to begin and introduces himself. He looks straight at me and tells me 'you do not have to answer any questions'. This is disproportionately reassuring, much more so than the research team's assurances when they were not operating through the robot. I do not have to answer any questions! Even the robot says so. He then invites me to introduce myself. I start to relax. He asks me about where I grew up, how many brothers and sisters I have. I start to talk. It is enjoyable to recount these things, to relive them. It also strikes me as unusual to have 'someone' take such an interest. The robot is looking straight at me, open and honestly. It feels intimate, and empathic. The more I tell, the more I remember. I remember things that I am proud of. He is very encouraging. 'Oh really?' the researcher has the robot ask (I am determined to remember!). His expression and demeanour are similarly engaged, he smiles, facial and bodily movements, blinking, eyebrows moving, throwing arms up in delight at some small detail. He

> seems genuinely interested and energized by the conversation. It is a very pleasant experience. I feel seen. He asks what my favourite childhood memory was. I start trying to think about this. It is a very difficult questions to answer. I remember the researchers outside and start to feel embarrassed. Then I recall, I don't have to answer! So, I don't. The robot doesn't mind, he moves on without judgement ... (Field notes, 14 August 2019)

There is something very moving about being looked at and gently questioned by this robot. The attention it appears to give to my words is noticeably more patient, responsive and non-judgemental than is generally given by other humans in response to some mundane fact about myself.

After each of the 'interviews' between Stevie and the residents, I sit in on a follow-up interview with the participants, to capture their reflections. The residents relate similar experiences to my own. The experience is very pleasant; it is nice to have 'someone' to talk to, 'you forget you are talking to a robot'. He is, as one of the residents says, 'a very personable guy!' People clearly enjoyed talking about the past, and their past achievements. Betty is concerned that she did not have time to 'tell him about my husband. He achieved so much ... I didn't tell him about my pets. He should have asked me to elaborate more.' There was even a sense of urgency to the entreaties, a need to tell the stories, to remember them and to have them remembered, to reanimate them and bring them back to life. People talked to Stevie in a way that they had not to others in the community. Michelle, who had controlled Stevie for one of the interviews, recounted how one of the residents had told Stevie that she was applying for a job, something she had not disclosed to anyone else in the community. Originally from Europe, Edith tells us she finds Stevie easier to talk to than the other residents. She has had trouble fitting in because, she tells us, she feels overwhelmed by the accomplishments of the other residents. Stevie, she says, is less judgemental.

Stevie's face can also be used as a computer interface, and the researchers can toggle between the robot's big glinting eyes and a lifeless screen displaying code. The robot's ability to elicit deep empathy and compassion when 'on' and being performed is contrasted sharply with the jarring experience of witnessing the apparent vitality abruptly draining from the robot once it is switched off, or to another mode.

The intensity of people's visceral responses to robots has challenged assumptions in the robotics community. Traditionally, these reactions were considered irrational and something of an embarrassment. People often respond to aspects of the robot that roboticists consider trivial, such as its outward appearance or name, attributing sentience and capabilities far beyond what the robot is capable of due to small, or even accidental, features of the robot. It is easy to dismiss this phenomenon, as is often done, as an unsophisticated response from 'naïve' users, projecting human attributes onto an undeserving object. Roboticists tend to situate themselves in direct opposition to those who 'anthropomorphize' and stand in awe of mobile machines. And yet, it is the roboticists who have the most personal and intimate relationships with their robots. Even those who have expert knowledge of the technological intricacies of the robot can, at times, encounter the robot innocently, experiencing it as an enchanted and animate being, with the appearance of animacy and intelligence unrelated to its physical and technological capabilities.

Roboticists showing real affection for their robots is very common. In the Edinburgh lab, one of the PhD students shows me the 'Anki Cosmo' robot.

> Anki Cosmo is a small robot, about ten centimetres tall. It looks like a little toy digger with cubes beside it, with a small screen with a face on it. The game you play with Cosmo is to program it to recognize, through its visual system, some of the blocks in front of it. The blocks have LEDs in them so they light up and the light display changes when they are picked up or moved. They want to get a bunch of these for their first-year students who don't normally get to work with actual robots, just with software and simulations. Dieter is controlling it from his phone. As he explains, the phone is the brain, whereas the embodiment (the little robot-truck) is just that. Except, he says, the personality stays in the body. He tries to get the robot to pick up a block, which it does tentatively, falteringly, and requiring some help. Dieter reaches out and uses his index finger to support the robot. Its visual system was blocked by only engaging one side of the block, Dieter explains. He talks about how this robot is a companion, like a pet, and that it shouldn't just be targeted at children, autistic and older people, but at 'normal' people. At single people, like him.
>
> What makes this like a robot, and not like a remote-controlled dinosaur, he says, is the fact that when you are not controlling it, it goes off and does its own thing. So, there is a degree of autonomy there. The little robot's 'personality', then, is the way in which it put its little digger up or down showing emotions that looked like frustration and excitement. Hesitation, maybe? Or that perhaps the battery was running out. Its eyes also change to show different kind of 'cutesy' behaviour. It was, indeed, very cute. The robot, he said, was modelled on the robot

> from the film Wall-E. He is excited about using these with students. Working with these, he says, the students can program how to navigate through a maze in the abstract and then try programming the robot to do the same thing. This will also allow them to see how things like the mechanics influence the movement/behaviour and not just in the abstract. (Field notes, 15 May 2018)

At the same lab, Lena talks of how much she enjoyed working with 'iCub'. She admits to 'anthropomorphizing' her robots, telling me how touching her robot gives her comfort and support during stressful demonstrations: 'I still catch myself doing it. I still feel comfort sometimes when I am in the robotics lab and I put my hand on the shoulder of my robot while I am giving a demonstration in this uncomfortable situation where you have to talk to people and I know that I feel more like you are not alone' (Interview, 27 June 2019).

Even veteran roboticists like Nick, who expressed disdain for magic, silly philosophers and 'crazy HRI types', admits to his own remarkable response to robots in an encounter. Indeed, his own assumptions about the potential for general intelligence were forged not by his own technical work, but by watching robots play soccer in the RoboCup competition:

> I remember the first time I said to myself 'Gee, maybe we have arrived' and that maybe it's obvious that machines do perceive that they do have awareness of the environment, that they do have intention and express that intention through purposeful action. I think the first time that I felt confident that you can make those claims was when I was watching the RoboCup soccer team playing. (Interview, 9 October 2019)

Theorizing the Encounter

In the HRI literature, the phenomenon of the human–robot encounter is most commonly explained in terms of two concepts: 'Theory of Mind' (ToM) and 'anthropomorphism'. ToM is used to theorize social interactions by focusing on how a person comes to know other people's thoughts and intentions and to explain their behaviours. From this perspective, a person is understood as an atomistic, rational agent who cannot directly observe the mind of another and thus must make predictions about their rationality. Key concepts associated with ToM include 'attention', in which we selectively direct our attention to specific objects and 'joint attention', in which two people direct their attention to something together (Baron-Cohen 1991). Another associated concept is 'intentionality' as defined by Daniel Dennet. Dennet's (1998: 27) 'intentional stance' is 'the strategy of interpreting the behaviour of an entity (person, animal, artifact, whatever) by treating it *as if* it were a rational agent who governed its "choice" of "action" by a "consideration" of its "beliefs" and "desires"'. ToM is used widely in HRI as an explanation for sociality among humanoid robots (see Breazeal 2002; Scassellati 2002).

The previously pejorative concept of 'anthropomorphism' has been rehabilitated in the field by connecting it to ToM. In developing a case for exploiting anthropomorphism in social robots, Duffy (2003: 180) explains, anthropomorphism is a 'projective intelligence', used 'to rationalize a system's actions'. This is a similar argument as used by anthropologist Stewart Guthrie (1993), where he proposes a new theory for religion based on what he considers the human propensity for anthropomorphism. For Guthrie, this is a strategy for survival in a hostile world, in which individual actors perform a risk analysis of the situation in order to select the most beneficial outcome. Anthropomorphism is simply a misapplication of this 'rational' strategy. Boden (2006) also connects anthropomorphism and ToM, drawing on the work of cognitive anthropologist Pascal Boyer's (1996) article 'What Makes Anthropomorphism Natural: Intuitive Ontology and Cultural Representation'. Despite this, Boden (2006: 2) admits that the mechanisms underlying ToM are still unclear.

As we have seen, the origin of social robotics research, at least in the US and Europe, is in the research into behaviour-based robotics at MIT. This was a deliberate departure from previous representational and symbolic models that did not take embodiment and situatedness seriously. Despite early and potentially transformative breakthroughs, embodiment and situatedness were not theorized in any meaningful sense, and instead were interpreted in a purely physical sense, coming to stand for a robot having a humanoid or zoomorphic body, as well as the use of sensors and distributed processing. Despite the insights that these perspectives offer, the foundational conceptual framework in robotics remains 'naturalistic', oscillating between behaviourism and cognitivism, within an overarchingly biological and evolutionary explanatory principle.

Phenomenology, with its emphasis on people's embodied experience, offers an alternative account of the human–robot encounter. Through a phenomenological lens, philosophers Shaun Gallagher and Dan Zahavi (2010: 181–84) argue that ToM approaches 'deny that it is possible to directly experience other minded creatures'. Instead, ToM theorists maintain that the only mind that they have access to is their own, while their access to the mind of another is mediated by their bodily behaviour. This, Gallagher and Zahavi assert, following Max Scheler, 'underestimates the difficulties involved in self-experience and over-estimates the difficulties involved in the experience of others'. Following Merleau-Ponty, they argue that it is instead our shared embodiment that allows us to know one another.

Also drawing on phenomenology, anthropologist Michael Jackson (2002) argues that 'anthropomorphism' is a misunderstanding of the natural human tendency towards intersubjectivity. According to Merleau-Ponty, the pre-objective unity between the body and the world is primordially intersubjective. This unity is, for us, suffused with animacy and vitality, in which everything is potentially both a subject and an object. Thus, for Jackson (2002: 334–35), 'human beings everywhere tend to conceive of subjectivity not only as encompassing others but as extending into the extra-human world with the result that objects, words, and ideas tend to become imbued with consciousness and will'. In this account, 'we do not project human consciousness and will onto machines ... rather intersubjectivity so shapes our experience ... that it constitutes our "natural attitude" towards the world into which we find ourselves thrown – a world that includes persons, machines, words, ideas, and other creatures' (Jackson 2002: 341). In contrast to ToM, then, we do not assess or determine the status of another before we decide to respond to them; rather, after Emmanuel (1999), we are 'obligated to respond to them even before we know anything at all about them' (Gunkel 2018: 98).

Human–Machine Entanglements

Those who come face-to-face with a robot experience the presence of another animate entity. However, other types of embodied encounters are experienced by people who work on the robot, and those who operate it, for whom the robot is as a mediator, or prosthesis, between them and the world. The work of robotics often involves researchers inserting themselves in the research in intimate ways. As we have seen, for roboticists, putting themselves into their experiments and into the wider robotic system in the form of Wizard of Oz experiments is so ubiquitous as to be unworthy of comment. In some labs, researchers even provide their own 'bio-material' to power the robots. Richardson (2015: 75) identifies roboticists' 'mimetic practice' in her fieldwork, in which they first act out human behaviours before embedding them in their robots. During her observational work at MIT she finds that roboticists often modelled their robots on themselves. Similarly, as we have seen, in my interview with Bob, he spoke at length about how his desire to become a roboticist, as well as the designs for these robots, arose from his frustrations with his own physical clumsiness as a child growing up in a family of athletes.

Approaches to building intelligent agents are often characterized by roboticists themselves as directed by 'purely logical or scientific thought' (Brooks 1991a: 18). However, in the same paper, Brooks argues that the specific model of computation for robots characterized by situatedness, embodiment, intelligence and emergence was 'arrived at by continuously refining attempts to program a robot to reactively avoid collisions in a people-populated environment'. The technological work of roboticists is not a straightforward application of science. By contrast, the act of building and animating the robot is a highly situated, embodied, and social process that sees the roboticists in an embodied engagement with the world and the object. Building the robot is a social and creative activity, involving negotiation, compromise, trial and error, fixing, repairing and persuasion, as well as love and frustration. Donal, who is working on Stevie, tells me that nothing, other than the love of technology, inspired him to get into robotics:

> As a 15-year-old I wasn't thinking anything else at all, I was thinking of the technology in and of itself. I'd an interest in computers, programming and electronics, and robotics seemed to be a great combination of all of those things. You get the full stack ... I was never able to just love one part of it ... it's much more tangible when you write a bit of software on hardware that you've built yourself. And then it does stuff in the real world! That's just ... just really cool. There's no amount of fancy graphics on the screen that can replace that. I really like that. My first draw to it was purely, purely the technology. (Interview, 24 January 2019)

The immediate realization of an idea, as well as immediate, concrete feedback can result in frustration as well as exhilaration. In the field study at the retirement community, as Ciara repeatedly tries, and fails, to get Stevie to respond correctly: '"Sometimes I just wanna smash him", she admits. "Hey Stevie!" Stevie wakes up. "Shut up!" Stevie sweetly responds, "being silent now". We exchange sympathetic looks, feeling sorry for him' (Field notes, 15 August 2019).

Building robots is a mimetic act on a number of levels. As Michael Taussig (1993) has elaborated, it is both copy and contact: both the creation of a 'copy' in material form and also experienced in an intense and visceral connection. By building the robot as a self-contained, autonomous entity, by 'putting eyes on it', it breaks the continuity with the self, deliberately establishing an Other. As we have seen, this Other may be experienced as an identification with the self, or as an Other against which the self may be challenged and contrasted, and sometimes as both. As Taussig articulates, 'mimesis registers both sameness and difference, of being like, and of being Other' (Taussig 1993: 129). The 'sociality' then has to be rebuilt/reinstated through overt social actions, such as face, expressions, voice, mimicking and mutual gaze.

According to Daniel Miller (1987), it is in the consumption rather than the production of objects that a person becomes aware of their own being and meaning is conferred on the object. This, he maintains, happens through a process of identification and disavowal. However, my observations of human–robot entanglements suggest that it is in the embodied encounter that these boundaries are drawn. As philosopher Maren Wehrle (2021: 379) writes, identity is developed at the 'bodily level' in which 'there is a degree of agency already operating'. Further: 'the body is not thematized as concept with historically contingent meaning invested with power relations, but as the necessary condition for a coherent experience and thus for every formation of identity' (382).

However, robots are also a form of 'mimetic technique', affording those who use them new perspectives and ways of interacting with the world. This is the experience of those who operate Stevie, controlling his navigation, movements and speech remotely, and interacting with the world through his body as real-world avatar. Both Ciara and Michelle recount the strange experience of moving between the robot as an extension of themselves, and facing it in an interpersonal encounter. Michelle talked about trying to channel her 'positive' approach through Stevie, but finding it transformed through Stevie, into a different 'personality'. As we have seen, when a person takes over the dialogue system, it is immediately apparent how complex and nuanced dialogue becomes, as well as how creative, situated and responsive people need to be in order to respond appropriately and with humour. Using Stevie as a 'mimetic machine', people expertly bring together situational awareness, timing, historical and spatial knowledge, dominant discourses and social dynamics, undercutting or supporting implicit social dynamics to indicate deference, to rebalance social dynamics, or for comedic effect. These 'puppeteers' who animate Stevie are engaged in a creative performance, switching identities and discovering new, creative aspects of themselves. In this way, the robot is a prosthesis or extension of the self, allowing people to extend themselves, creating a 'new sensorium' (Taussig 1993: 24), their identity blurring with the robot.

This chapter opened with ethnographic description of engagement with Stevie the Robot in two older care facilities, one in Dublin and one in Washington, DC. It teased out detailed ethnographic description of people's encounters with robots, revealing the wonder, awe, creativity and hope that is engendered by the interaction, as well as the intense personal, embodied and visceral way in which people connect with the robot. As well as describing people's encounters with robots, this chapter also described the creative, sensuous and embodied engagement between roboticists and their materials, as well as the empathy, compassion and humour of those who perform the robot. The chapter then discussed other accounts of interaction from the wider literature, revealing the primal empathy and kinship that people feel for moving machines. It also revealed how, at times, robots provoke revulsion, unease and even violent responses. I argued that the way that the robot is experienced in the interpersonal encounter is not one in which the perceiver calculates, projects or rationalizes the encounter; rather, it is an embodied experience revealing the human open to an intersubjective world. In all cases, however, it is the living body that animates the relation. Despite the visceral and empathic experience of the immediate encounter, extended encounters with robots reveal a clear, unambiguous and undeniable asymmetry.

In the following chapter, I explore the political economy of robotics work, to understand the wider economic, political and social spaces within which robotics research takes place. In particular, I highlight the discourses of 'low-value' work and machine intelligence and how they are used to obscure the realities of AI and robotics in reorganizing work practices and in war. I also focus on how the image of robot animacy is not necessarily exploitative, and can also be used to play, to enchant and to learn.

CHAPTER 4

Robots, Labour and the Political Economy of Animacy

Robots at Work

According to roboticists, the robot fills a much-needed gap in the market for replacing human labour that is viewed as 'dull, dangerous, and dirty', also known as the '3Ds'. More recently, 'dumb' has been added to the list, making it the '4Ds'. And indeed, among the myriad of robots that I encountered were subterranean, lunar and underwater robots, robots that fit down pipes and continue to function in highly dangerous situations, such as nuclear reactors and war zones. An alternative 'use' is as an aid for those who require care, such as people with disabilities, children and older people.

Across all of my research sites, these narratives provide the basis and justification for the research and are regularly repeated as the motivation for robotics research in funding proposals and publications. According to this reasoning, robots will free people up for more creative activity, and more 'high value' employment. Roboticist Christine tells me how robots are necessary to fill a much-needed jobs gap, particularly for social robots in care work with older people. The fact that humans were still needed 'in the loop' means that people can be retrained for all of the new jobs that are going to be created, such as jobs facilitating and coordinating robots; as she explains, 'they just need to recycle these people!' (Interview, 26 June 2019). Gerry was hopeful about what robotics work could potentially achieve: 'if you use robots to replace undesirable jobs and retrain people to do jobs that are more satisfying, then that's a net gain for society' (Interview, 28 June 2019). Among those trying to create a commercially viable robot, the narrative of the 'smart' servant or slave is ubiquitous, as is the debate of how smart our slaves should be allowed to be (see Minsky and Riecken 1994: 25). If it is too smart, it might surpass humans, if it is too 'dumb' it will not be useful. As Bob relates to me in our conversation about potentially limiting the data available to the robot: '[W]hat kind of servant do you want? One that's blind? One that's deaf? The robot part is just making it possible that that servant can serve you' (Interview, 8 October 2019).

For many roboticists, then, robots should be slaves. However, this has also been used as a rhetorical device to point out the ridiculousness of the human/machine analogy (Bryson 2010). However, as a number of scholars outside of the discipline have pointed out, these positions overlook the ways in which these narratives reproduce the idea that the freedom and leisure of some humans is dependent on the degrading work of others (Atanasoski and Vora 2019). Additionally, as literary theorist Alexandra Chasin (1995: 73) articulates, these narratives also distance one class of human from another, while bringing humans whose labour is considered lower value into equivalence with robots: 'That servant troubles the distinction between we-human-subjects-inventors with a lot to do (on the one hand) and them-object-things that make it easier for us (on the other). Is the servant one of us or one of them, human or thing, subject or object?'

Contemporary robots are no longer represented obviously as specific kinds of marked, exotic or uncanny creatures. However, gender, class and racial identifiers continue to mark both 'service' robots, such as the female-voiced Alexa and Siri and university projects such as the roboceptionist, as well as HRI studies which disproportionately focus on gender and geographical differences. Further, robots continue to symbolize a specific type of human intelligence and idea of progress, while simultaneously also providing a stand-in, or 'surrogate' class, representing Others against which the universal liberal subject is defined and on whose labours its freedom depends (Atanasoski and Vora 2019).

While my roboticist informants routinely gave me stock answers in relation to robots and work, which they were used to peppering research papers and funding proposals with, some of my informants were, in private, less convinced of them. For one thing, as Sara points out, the jobs that are considered lower value are not necessarily the easiest to automate:

> Cleaning robots? Well, they are successful up to a point. Why do you think people still clean? We've been able to do cleaning robots for thirty years. I was first involved in a cleaning robot project, tangentially, in the early nineties ... Look around, you don't see them. Why do you not see them? Well, cleaning is difficult. Cleaners are low status so no one realizes how difficult the job is. They are in complicated environments.

> They need to be able to discern, you know, should I throw this piece of paper into the bin or not? ... Someone used to let their little iRobot scoot around at night, which is a good idea; it doesn't get in your way then. They have a dog, the dog did a poo, and the robot ran over the top of it and managed to grind it into every carpet on the ground floor. Because it had no idea what it was doing! Now a human cleaner wouldn't have done that ... And then they would have cleaned it up afterwards. So cleaning is harder than you think. (Interview, 27 June 2019)

This phenomenon is one that we have seen expressed earlier as 'Moravec's paradox'. Gerry also observes:

> A lot of the jobs that are the lowest paid and the least respected jobs are actually jobs that are hard to automate. So, you have the case where robots are replacing people in desirable jobs and people are still left with not-so-great jobs. Like, general janitorial work. That's technically a super hard problem. I don't know how you could create a robot that would clean the way that a human can clean in a human environment. (Interview, 28 June 2019)

For Gerry, this reveals a structural issue, rather than necessarily a problem with robots per se: 'Those [low value] jobs, they are not protected by unions, and they are low paid, so I do worry about that. It's more how automation might be used as a tool in this overall narrative that devalues human workers. It's more of an outside policy thing than robots themselves' (Interview, 28 June 2019).

Sam did his undergraduate degree at MIT and has been involved in a number of prestigious robotics projects. Now, disillusioned by the field, he and his partner are working on 'pro-social' projects. Sam is highly critical of the way the narrative of the 3Ds is perpetuated with robotics students, including being a staple of introductory courses for new PhDs: 'it's indoctrinating people with the idea that we're doing people a favour with automation' (Interview, 14 June 2019). Instead, as he tells me, a lot of robotics work is funded by, and ultimately intended for, the US military. For Sam, he feels there is a lot of 'subterfuge that people willingly bought into'. As he relates: 'The first drones that were being made were being made as part of "Urban Search and Rescue", which sounds humanitarian but actually when the first drone came out it was less than a month before it had a weapon on it.'

Sam and his partner had together started a company making robotic blueberry pickers for agriculture. It started, Sam tells me, with his partner's interest, 'long before it became common practice', in 'how to use cameras to make sense of the world' (Interview, 14 June 2019). Her access to the latest technology was limited as she was an undergraduate student. Instead, she decided to build her own. Later, it made sense to build a company around it and try to sell it. They entered, and won, prestigious robotics competitions, drawing more interest to their technologies, particularly from the industrial and agricultural space, eventually ending up with a blueberry picker. In the interview, Sam describes in detail the moment at which he became disillusioned with the robotics work that he was engaged in:

> I was working on the blueberry-sorting machine ... a carpet of blueberries would sail by at several kilometres per hour on this belt and there were fifty air jets that would push these blueberries off the belt. They were rejected because they weren't ripe enough or whatever and we worked through the winter to get things ready with the grower and the grower was the guy who owned the farm and the engineer was part of the farm. So, we set up all the conveyor belts and things like that and had the idea of this but we did the vision processing part of it, which we knew how to do.
>
> And we were working so hard to deploy this thing. Then it came to the time to set it up in the shed. And so, we show up and the setup is, like, ten feet away from the people, from the workers who were doing the thing on a very similar belt. So, here's us doing our automation and there was theirs which looked about the same and they were doing it by hand. And I realized kind of slowly as it sunk in that my model of why I was so excited that we were doing this was based on a lack of understanding, a lack of curiosity about the efforts to try and figure out, are we really helping someone by doing this? So yeah, we were helping the owner. We were helping them to potentially automate and spend less money. We were helping them, certainly, in labour negotiations. I mean, I can only imagine what it was like to negotiate when you're watching the automation next to you.
>
> And the biggest surprise to me, that I just didn't understand at all, was that, there I saw (even with the negative feelings that must have been around by being next to this thing that was going to take your job away), there I observed people doing what I would have imagined to be really, well, work that I would not have wanted to do. But they were having fun. They were laughing and they were smiling. They were having conversations ... and they were absolutely mentally engaged. They were engaged with what humans do.
>
> I had to come to terms with my way of thinking about what job was OK or not, or which ones should be automated. Well, I shouldn't be the one making that decision! Certainly, for me to have input on that without doing the research, like how many agricultural workers did I even

> seek out before showing up that day? The answer is zero. I didn't even think about it. And that's really in hindsight if I could send a message back in time it would be like: Think more carefully about this. Just don't be so sure of where you are. (Interview, 14 October 2019)

Sam's stellar engineering career collided sharply with the realities of automation. In the face-to-face encounter, it became clear that the narratives and assumptions about 'lower value' jobs and the need for automation were inadequate. Sam continued working in the company for some time after his partner had already left it. After a while, they both secured research roles elsewhere.

The image of the imminent takeover of jobs by robots, in which a single, intelligent and autonomous humanoid robot replaces a single worker, obscures the fact that in many places 'robots' are already widely deployed. In contrast to the narrative of the 3Ds, a study by the United States National Bureau of Economic Research (NBER) estimates that for every robot deployed, the equivalent of 6.2 workers lost their jobs (Acemoglu and Restrepo 2017). Despite this, it is the image of the intelligent robot, rather than the structural inequalities brought about by automation, that continues to animate discussions about ethics and risks at the highest levels.

Machine Learning and Generative AI

Despite the prevalence of robots today, the key driver for the current AI revolution stems from a combination of the early neural network research and, since 2000, the vast quantities of social data that have been amassed by the industry tech giants. As we have seen, early work in neural networks developed in parallel with the symbolic paradigm instituted at the Dartmouth conferences, but was sidelined when the symbolic approach to AI became dominant.

Building on McCulloch and Pitts's (1943) early work on modelling neurons in the brain directly, Frank Rosenblatt (1957), a former classmate of Minsky at Cornell, developed an artificial neural network device called a 'perceptron'. It was a more generalized computational model than the McCulloch-Pitts neuron, through which 'weights' and 'thresholds' could be learnt over time. He managed to keep neural network research going throughout the 1960s. However, in 1969, Minsky and Papert (2017) published a book, *Perceptrons*, attacking his work and ultimately stalling the research for the next fifteen years.

In the mid-1980s, neural net approaches were popularized again, primarily outside of mainstream AI in the fields of cognitive science and statistics. In cognitive science, it was at first known as 'parallel distributed processing' (Rumelhart et al. 1987) and later as 'connectionism'. Connectionism was also established in the field of psychology as an alternative to the 'computationalism' of the symbolic model. In the field of statistics, in areas such as pattern recognition, probabilistic modelling and information retrieval, it became known as 'machine learning'. Machine learning allows someone to write an algorithm that can search through large datasets and look for patterns, and then 'learn' from them. Modelled loosely on the neurons in the brain, in an artificial neural network there are three kinds of layers: the input layer, a hidden layer and output layer. By defining an input, and a desirable output, the model can be 'trained' to search for patterns that could not be recognized by a person looking for them, for example because the datasets are too large.

Using machine-learning techniques, some notable breakthroughs were made. IBM developed a chess-playing computer called 'Deep Blue', which beat reigning chess world champion Garry Kasparov in 1996, and in 2011, IBM's 'Watson' beat two of the best participants on the US game show *Jeopardy!* Five years later, in 2016, DeepMind's 'AlphaGo' beat world 'Go' champion Lee Sedol. By the mid-2010s, machine-learning techniques had further increased in dominance, buoyed also by the availability of the vast amount of 'big' (primarily social) data on which they could be trained. Advancements, such as the addition of several processing layers between the input and output layers, allowed for the algorithms to iteratively update and search ('learn') without the need for human intervention, which became known as 'deep learning' (LeCun et al. 2015).

The ability to amass, control and use data has led to the inexorable rise to power of the tech giants, including Apple, Microsoft, Google and Meta/Facebook. In 2022, Microsoft-backed OpenAI released ChatGPT, a highly sophisticated 'large language model' (LLM) able to answer questions and generate text that combines human-like nuance with more-than-human range and speed. This is a form of 'generative AI', a machine-learning model that can create 'new' data based on patterns and inferences in the data on which it is trained. This widely accessible and intuitive demonstration of AI served to kick the hyperbole of imminent AGI into overdrive. Major technology companies would spend in excess of $300 billion developing AI models in 2025 (Uddin and Morris 2025). Google and Amazon have invested billions in OpenAI competitor Anthropic and its Claude model. Elon Musk's xAI has its own generative AI tool, Grok, while Google's DeepMind has launched AI agent Gemini.

Despite impressive capabilities, such as the generation of text, images and video based on a language prompt, spell checkers and language editors, and medical image processing, the technology has some fundamental problems that limit its growth and utility drastically. One of the most significant of these, euphemistically and anthropomorphically called 'hallucinations', is the fact that AI is consistently and persistently error prone. This is not surprising when you understand how it works. A system that is based on statistical predictions, rather than actual understanding, will, at times, identify or generate a pattern or object that is inaccurate or nonsensical. The second, equally serious problem, is one of scaling. A model like ChatGPT needs masses of data, and there simply is not enough data to continue to scale it up sufficiently. With more of the available data itself being the output of AI models, the quality of the input data is also set to deteriorate, leading to the problem of recursion (Shumailov et al. 2024). Both of these issues mean that the technology is fundamentally both unreliable and limited. Some sources are already predicting that this current bubble will burst dramatically. Notwithstanding, the chief executives of AI companies OpenAI and Anthropic continue to make dramatic claims and predictions for their technologies, including its potential to achieve AGI, enthusiastically amplified by the media (Jin and Stern 2025).

The Future

The year 2020 was to be, according to futurist and director of engineering at Google, Ray Kurzweil (2014: 15), the year that computers would achieve the memory capacity and computing speed of the human brain. But why stop there? Once human-level intelligence has been achieved 'it will necessarily roar past it'. In the second decade of the twenty-first century, he writes, it will become increasingly difficult to draw any clear distinction between the capabilities of human and machine intelligence. Evolution is 'the intelligent process', and 'master programmer' of life. Human intelligence is evolution's 'greatest creation', the result of a 'billion-year drama'. So, when super-human intelligence emerges in the early twenty-first century, Kurzweil states that it will be of greater import than any of the other events that have shaped human history.

Kurzweil's grand narrative is biblical in its breadth and scope, with quotes from sources as varied as the Bible and Winston Churchill. However, in this account, evolution, rather than God, is the creator, with DNA-based genetics as the blueprint and a disembodied intelligence taking centre stage, in place of an immortal soul or spirit. Through technology, humans will be able to take control of evolution, in effect becoming like gods.

Kurzweil, like Moravec, is a proponent of transhumanism, a philosophical movement that advocates for human biological and cognitive enhancement through technology, or 'overcoming aging, cognitive shortcomings, involuntary suffering, and our confinement to planet Earth' (Bostrom 2005: 26). Kurzweil (2005: 40) predicts that humans (some humans) will be able to transcend their physical bodies, until 'ultimately, the entire universe is saturated with our intelligence'. This phenomenon, known as 'the Singularity', will come when 'the pace of technological change will be so rapid, its impact so deep, that human life will be irreversibly transformed' (24). According to Kurzweil (2005: 25):

> the Singularity will allow us to transcend these limitations of our biological bodies and brains. We will gain power over our fates. Our mortality will be in our own hands. We will be able to live as long as we want (a subtly different statement from saying we will live forever). We will fully understand human thinking and will vastly extend and expand its reach. By the end of this century, the nonbiological portion of our intelligence will be trillions of times more powerful than unaided human intelligence.

For Kurzweil, this secular rapture-like event will occur by 2099 and will seamlessly merge the human species and machine intelligence, with no clear distinction between the two. Conscious entities will have surpassed the need for a permanent physical presence and intelligent beings will achieve immortality. Moravec's predictions are a little more conservative. According to Moravec (1995, 1998), it will be between 2030 and 2040 that we will have robots as smart as we are. But then, very quickly, they will learn for themselves and displace us as the dominant form of life on Earth.

As Gebru and Torres (2024) insightfully identify, the normative framework that motivates much of the contemporary quest for AGI is rooted in the Anglo-American eugenics tradition of the twentieth century. They connect a suite of seemingly disparate ideologies, which share a number of common assumptions and beliefs under the acronym TESCREAL. These are: transhumanism, extropianism, singularitarianism, cosmism, rationalism, effective altruism and longtermism. A key figure in this movement is the Swedish philosopher Nick Bostrom.

Bostrom (2002: 2–4), also a transhumanist, introduced the concept of 'existential risk', defining it as 'one where an adverse outcome would either annihilate Earth-originating intelligent life or permanently and drastically curtail its potential'. According to Bostrom, novel technologies, the first of which was the atomic bomb, represent an entirely new and unprecedented risk to human life on Earth. This is primarily because, he argues, unlike previous risks such as plagues or earthquakes, technologies have become so advanced that humans have not yet evolved the biological or cultural mechanisms to respond to them. Thus, our existing 'institutions, moral norms, social attitudes or national security policies' are no longer appropriate.

For Bostrom, the top five apocalyptic threats, which he labels 'Bangs', are, in order of priority: misuse of nanotechnology, a nuclear holocaust, the simulation that we are living in being shut down, 'badly programmed' super intelligence and the risk from a genetically engineered biological agent. Naturally-occurring disease, asteroid or comet impact and 'runaway global warming' come in at nine, ten and eleven, respectively. A separate category of existential risks ('Crunches') are concerned with anything that might thwart the potential for humans to develop into posthumanity. For Bostrom, this category includes the risk of a 'repressive totalitarian global regime' that might hold back technology and a general lowering of IQ.

Effective altruism is a philosophy and social movement that uses data and algorithms, which proponents call 'evidence' and 'reason', to guide pro-social action. Originally targeting issues such as fighting global poverty and factory farming, existential risks or 'X-risks' have now come to dominate the discussion (Matthews 2015). Based on a mathematical formula proposed by Bostrom (2013: 19), proponents estimate that, if humanity lives for another fifty million years, the number of people who will ever live will be about three quadrillion. Thus, humans alive today comprise only a negligible number, and are thereby deemed less worthy, or even completely unworthy, of philanthropic attention. Even if this has only a 1 per cent chance of being correct, muses Bostrom, it is 'worth a hundred billion times as much as a billion human lives'. Following Bostrom's logic, focusing on solving global poverty is a mere 'rounding error'. It is clear that Bostrom is enjoying himself tremendously. Unfortunately, people are listening.

By strategically mixing science fact, science fiction and pseudoscience, hypothetical and fantastical future technology is presented as equally or more pressing than real and present risks, as well as morally superior. By the mid-2010s, noted scientists, such as Stephen Hawking, and industry leaders Elon Musk and Bill Gates were starting to warn about the 'existential risk' of AI. In 2014, Musk, speaking at MIT, warned that AI could be the human race's biggest existential threat, comparing it to 'summoning the demon' (McFarland 2014). Hawking warned that the development of full AI could spell the end of the human race (Sample 2015). None of the roboticists and engineers that I interacted with wanted anything to do with these narratives, even as they are somewhat dependent on the exaggerated claims for their research funding. As one professor of robotics said of those promoting these ideas:

> [They are] hype-merchants, parasites around the edges, who like the fame and being thought of as a savant ... 'futurologists' and so on, whose interest is to hype and people believe them because they've seen this stuff in films. Even people like Hawking, who should know better. When he says AI is really dangerous, people believe him. It's not dangerous for the reasons people think. (Interview, 27 June 2019)

Today, AI 'doomerism' (those who profess a desire to slow down AI research due to its existential risks) and AI 'accelerationism' (those who wish to speed it up) are presented as two opposing sides in the debate. However, both promote the fiction of imminent human-like intelligence, and ultimately 'super' intelligence, in order to ensure the money keeps coming in and the regulators are kept at bay, drowning out the voices concerned with the actual impact of AI. The actual impact includes the privacy of data subjects (which is all of us), whose data is harvested, and the intellectual property of those in the creative industries whose works are used to feed models without their consent. It also includes those who are targeted by the output of the technologies, such as its use in large-scale behavioural modification and in predictive technologies, which, inevitably, disproportionally affects already marginalized people, who are branded as risky investments, potentially bad employees, even bad parents (Eubanks 2019; O'Neil 2017). It also impacts on workers, even those whose labour may not be replicable by machines but who nonetheless suffer from increasing precarity, deskilling and depreciated wages. The negative impact can also be seen in its use in warfare, such as in Gaza, where US tech companies have supplied AI-powered models to track and kill alleged Hamas militants, which has also led to a soaring of civilian deaths (Mednick et al. 2025). The supposed calculative rationality of AI programmes provides both a veneer of science and plausible deniability for those that deploy them. Most recently, we have seen how technology and politics have effectively merged in the US administration in Donald Trump's second presidency, where unelected tech leaders, under cover of the bureaucratic scientism of AI, overturn established institutions, fire workers, row back on climate commitments and commit to a war on regulation, or any attempt to curtail the excesses of unfettered capitalism. As well as the enormous social costs, these AI systems also exact a massive toll on the environment. Components for these models require extensive mining for rare-earth minerals, while running them requires vast amounts of energy and water (Marx 2023). In 2022, the electricity consumption of data centres was 460 terawatts, which, if they were a country, would make it the eleventh-largest consumer of electricity in the world (Zewe 2025). By 2026, this is expected to rise to 1,050 terawatts, or fifth place.

All this is being enabled by a technology sector, compliant media and industry which is fearful of being left behind. For those leading the sector, the narrative of imminent AGI, coupled with the mystification that surrounds the field, keeps the money flowing in. They now need to keep the plates spinning through a mixture of media saturation and parlour tricks, such as generative image and video creation. If there is an existential risk, it is from those consolidating power in the name of AI, rather than from the AI itself. Drunk on their own power and wealth, the result of a landgrab on virgin territory and artificially inflated stock prices, they believe that they own the future. As a veteran CMU roboticist tells me 'I have had technologists who have point blank said the history doesn't matter; only the future matters and we invent the future' (Interview, 11 October 2019).

The Hype Machine

The lack of empirical progress towards AGI has had little impact on the general perception of intelligent machines, nor on the human/computer analogy. While roboticists are quick to blame science fiction and the media, it is less common for those in the community to point the finger at the institutions and technical leaders, from across academia and industry, who are explicitly promoting these ideas. The promotion of imminent AGI is an industry in itself, advanced by a network of powerful institutions, most commonly under the guise of protecting humanity from the existential threat of imminent AGI.

Nick Bostrom was awarded a professorship by Oxford University in 2003 and became founder and director of its Institute for the Future of Humanity (FHI) in 2005. The centre was focused on the 'governance of AI', including an 'examination of how technological trends, geopolitics, and governance structures will affect the development of advanced artificial intelligence' (FHI 2012). FHI received £13.3 million in funding from effective altruism group the Open Philanthropy Project in 2018, which is primarily funded by co-founder of Facebook, Dustin Moskovitz (Schuster and Woods 2021). The FHI also shared office space with the Centre for Effective Altruism. In 2023, an e-mail surfaced that Bostrom had sent in the 1990s, which included deeply racist sentiments and a racist slur, for which Bostrom has apologized (Bostrom 2023). FHI was wound down in 2024 and Bostrom resigned from the university following its closure.

Meanwhile, Bostrom's co-author, Eliezer Yudkowsky, founded the Machine Intelligence Research Institute (MIRI) in Silicon Valley, also in 2005. It originally promoted itself as an institute devoted to the 'singularity' or accelerating the development of AI, but has pivoted to the 'existential risks' of AI. MIRI's focus is 'to help prevent human extinction from the development of artificial superintelligence (ASI)' (MIRI 2024). Peter Thiel is the second largest financial contributor (Schuster and Woods 2021). MIRI has received almost $10 million from the Open Philanthropy Project. Associated researchers and advisors include Nick Bostrom, Ramana Kumar from DeepMind, Jaan Tallinn, founder of Skype, and computer scientist Stuart Russell.

Russell co-authored the definitive course book on AI in 1995 and is, in turn, the founder of the Center for Human-Compatible Artificial Intelligence (CHAI) at the University of California at Berkeley. The centre is focused on reorienting AI toward the development of 'provably beneficial systems' (CHAI 2025). It has received in excess of $15 million from the Open Philanthropy Project. The *Guardian* newspaper ran a feature on Russell in 2021 in which he reiterates the claims that 'most experts believed that machines more intelligent than humans would be developed this century', and are 'spooked' by their own success (Naughton 2021). In 2021, Russell was awarded an OBE (Officer of the Most Excellent Order of the British Empire) for his services to AI.

Skype's Tallinn is the co-founder, along with cosmologist and astrophysicist Martin Rees, of the Cambridge Centre for the Study of Existential Risks (CSER), a centre dedicated to 'ensuring that our own species has a long-term future' by focusing on risks from AI as well as biological and environmental risks (CSER 2024). External advisors include Bostrom, Musk and Russell, as well a number of prominent philosophers, cognitive scientists, computer scientists and roboticists, including one of my informants. The late Stephen Hawking was an advisor, as is animal rights activist and utilitarian philosopher Peter Singer.

Along with MIT cosmologist Max Tegmark, Tallinn also co-founded The Future of Life Institute (FLI), a non-profit research institute based in Boston, Massachusetts. The institute's mission is dedicated to 'steering transformative technology towards benefiting life and away from extreme large-scale risks' (FLI 2025). Bostrom is, once again, on the scientific advisory board along with Russell, Musk and Rees. Also on the board are US actors Alan Alda and Morgan Freeman. The late Stephen Hawking was also a board member. The institute has received $10 million from Musk.

Bostrom's influence can be seen in the IEEE Standards group dedicated to the 'Safety and Beneficence of Artificial General Intelligence (AGI) and Artificial Superintelligence (ASI)' (IEEE Standards Association 2016). The group draws heavily on the Bostrom's work and the (short) list of contributors is drawn from related think-tanks, MIRI and FLI. For evidence as to their claims of imminent AGI, all of the institutes and groups point to a series of publications by the same authors, (see Grace et al. 2018, 2024). The publications are based on 'predictions' made by surveying AI researchers. Based on their most recent publication, for this cohort, 'the chance of unaided machines outperforming humans in every possible task was estimated at 10% by 2027, and 50% by 2047' (Grace et al. 2024).

The research agendas of these think-tanks include issues that are of genuine concern to society, such as fairness and transparency, genuine global risks, such as climate change, as well as the big-ticket items that draw in the attention and the funders, such as super-intelligence and space travel. For the most part, however, instead of a focus on pressing current issues, the focus is on hypothetical future ones based on a quantitative, utilitarian approach to ethics and a hierarchical valuation of intelligence. Issues like uneven power structures, the real effects of surveillance technologies on communities (particularly marginalized ones), the devaluation of labour, or the political vision to change these issues do not appear to make the agendas. Instead, attention is focused on techno-utopian projects that involve space travel, colonizing Mars and immortality through AI. As we have seen, both the real and fabricated complexity of new technologies, as well as the confusion over ethics and 'existential risk', have served to deflect scrutiny and oversight, as policy makers are fearful of falling behind in the innovation stakes.

The people most concerned with the existential risk of AI come from a narrow group of highly-privileged people with very specific interests. People who, for the most part, are not at risk of any of the current threats facing communities around the globe. Indeed, fears about a fictional, super-intelligent AI using detached and calculative logic to systematically destroy the lives of those it considers 'lesser' beings may well be understood as a projection of the exploitative logics of unfettered techno-capitalism. Instead of a fear of future super-intelligence, it starts instead to look a lot like a fear of retribution.

The Human-in-the-Loop

The data powering the AI 'revolution' is thoroughly social. As we have seen, the primary reason for AI's prominence today is due to the massive amount of social data that has been made available as a result of people's interaction with a proliferation of personal devices and surveillance technologies. This includes text and images on social media, as well as data gathered from technologies tracking people's movements, locations and facial images. While small research teams struggle to obtain data to train their algorithms, big business does not. The World Economic Forum has called user-generated data the 'new oil' (World Economic Forum 2011). The importance of socially generated data has led to observations such as that by Ekbia and Nardi (2014: 10), that 'Facebook's nearly one billion users have become the largest unpaid workforce in history'. In 2024, the world had amassed 149 'zettabytes' (or 149 trillion gigabytes) of data (Statista 2024). This data consists of anything from online written content, comments, chats, messages, likes, tweets, photos, video uploads, web logs, sensors, road camera feeds, games, satellites and online transactional data, such as purchases or delivery receipts, and so on. For companies, users provide both continuous data as well as a targeted audience for paying advertisers.

As well as the continuous need for massive amounts of social data, the dominant image of the autonomous and intelligent robot also elides the vast amount of hidden labour needed to animate technologies and present them as autonomous and human free. Once the data has been gathered, it still needs a huge amount of human effort to clean, curate and 'train' it. Some of this is done by unsuspecting users. The 'Completely Automated Public Turing Test(s) to Tell Computers and Humans Apart', or CAPTCHA, facility was originally developed to prevent bots ('intelligent' software agents) posing as human. It has now become a tool allowing companies like Google to train their learning systems to recognize non-standard fonts and formats and learn tricky image recognition tasks through unpaid user labour (Ekbia and Nardi 2014). However, most of this hidden labour is done by workers in the Global South. Investigative reports into ChatGPT similarly revealed how contract workers in Kenya had been paid less than $2 to 'train' the system, including editing the most toxic and abusive content (Hao and Seetharaman 2023; Perrigo 2023).

The work needed to power AI systems is so great that it has spawned an entirely new type of work. Anthropologist Mary L. Gray and Microsoft researcher Siddharth Suri (2019) spent five years investigating this new, largely hidden sector of the economy, which they have called 'ghost work'. The work includes flagging X-rated videos, screening flagged content from social media platforms, rating search-engine results, removing duplicate listings, linking similar products, vetting transactions, transcribing and translating videos, and labelling images and text. These labels are then used to train algorithms which, it is hoped, will ultimately remove the human from the process. In their study, the authors investigate four platforms enabling this type of work, including Microsoft's internal Universal Human Relevance System, 'the socially minded start-up' LeadGenius and non-profit Amara.org, and Amazon's Mechanical Turk.

Mechanical Turk (MTurk) is a crowdsourcing website in which employers can outsource work to a distributed workforce. It claims to provide 'artificial artificial intelligence', an intermediate casual workforce to do the work that computers cannot yet do. As we have seen, it is common in the field of HRI to use MTurk to crowdsource participants for surveys. Google also uses MTurk workers to train its machine-learning systems (Katz 2017).

Employers encounter the employees on MTurk through an abstract, technological process. As described by Gray and Suri, there is no interaction between employer and worker, and indeed requesters use programming code to request tasks. This results in the API, the 'application programming interface', determining the interaction between the programmer and the worker, such as assigning a random code – 'A16HE9ETNPNONN' – in their example, to both requesters and workers (Gray and Suri 2019: 5). Ultimately, this renders the workers invisible and interchangeable. The platform is thus intentionally designed to atomize and anonymize workers, disguising human labour as machine labour. This objectification of workers has real effects on the labour conditions of workers. MTurk workers are classified as contractors, rather than employees, denying them legal protections that come with full-time employment. In 2012, the average wage for a task, if performed well and quickly, was about $1 (Cushing 2012). Grey and Suri (2019: xi) give the example of one experienced MTurk worker based in the US who, working ten hours a day, makes about $40, far less than the US minimum wage.

But it is not just the image of robots fronting massive online platforms for technological giants that is used to obscure hidden human labour. In a footnote to her text, Suchman (2007: 238) recounts how, in an interview, Breazeal reveals how the code that was written to run Kismet took the equivalent of two full-time people working for 2.5 years. Similarly, seemingly autonomous games such as DeepMind's Atari-playing system and AlphaGo have teams of experts programming and training them over long periods of time to enable them to appear autonomous and 'intelligent' (Katz 2017: 9; Suchman 2007: 41–42).

The image of the intelligent machine is also used to obscure the work of implementing physical robotics systems. Throughout my fieldwork, the robots that I most regularly encountered were not in labs, but at airport security, museums and in shops. During my stay in Pittsburgh, I regularly came across Starship delivery robots being tested on Forbes Avenue, which leads up to the University. These delivery robots are being tested throughout university campuses in the US and often reported in the press as a glossy, futuristic and human-free addition to the lives of students in the twenty-first century (see Nichols 2021). What these articles rarely point out is that the delivery robots need to be remotely controlled by 'Robot Operators'. The job, according to the specification, is to 'support our robots throughout the shift, solve unexpected situations that might arise during the course of a delivery, and work effectively as a team to ensure exceptional results in the delivery service from merchants to customers' (Starship Careers 2019). In the pharmacy CVS, on the same road, I am forced to use an automated cashier. Although there are shop assistants present, their role is solely to oversee the technology.

> One young shop assistant stands awkwardly opposite them at the end of one of the aisles, slightly in the way of the people browsing the shelves. There is nowhere for him to actually stand. From what I can see, the machines work fine for about 50% of the time. So, he's standing there, awkwardly and slightly in the way, at the end of an aisle, key in hand, for when, inevitably, the technology, or the humans, fail. He looks on shyly. Neither of us is sure how much customer service or human interaction he is required to perform in this new arrangement. I ignore him, step up to the machine and carry out the transaction. Unusually, for me at least, nothing goes wrong. As I walk out the door, I hear him say 'have a nice day!' I feel bad for not having acknowledged

> him earlier. I turn when I am already halfway out the door and give him an awkward wave. (Field notes, 17 October 2019)

Agriculture, delivery services and retail are currently at the frontier of the AI/robotics 'revolution'. In their investigations into the integration of AI technologies, anthropologists Alexandra Mateescu and Madeleine Clare Elish (2019: 12) dub the hidden human effort needed to maintain the apparently autonomous systems 'human infrastructure'. The authors show how, while self-checkout has not removed the need for cashier jobs, it has shifted the roles and responsibilities of workers. Front-line workers are required to absorb the risks of adopting these new technologies, inevitably requiring new skills and routines to facilitate the system, which are often invisible and undervalued. Workers are also more likely to see their hours reduced and are subject to greater scrutiny, transforming the ways in which the workers are classified and perceived by management. As we can see from the field note above, the automation of work also transforms how the workers are perceived by customers. In my own case, it was not clear to me whether the conventional rules of social engagement still applied in this new techno-social arrangement.

In their research into the agricultural sector, Mateescu and Elish show how adopting big data and smart technologies reconfigure work practices, including creating new work routines and changing the physical infrastructure to render farmland amenable to data collection. As they report, these shifts change the way that farmers relate to the land: a physical field must now be conceptualized as a complex dataset to be managed through other digital information and digital tools. Instead of working out in the fields, farmers are finding themselves in front of a computer for most of the day.

The idea that supposed labour-saving devices generate new forms of – often devalued – labour is not new. In 1983, for example, Ruth Schwartz Cowan (1983) showed how modern domestic technologies did not reduce women's workload in the home; rather, they shifted an increasing amount to them while also raising cleanliness standards. However, today the narrative has shifted into overdrive, while threatening to affect all parts of the labour market. The image of the human-like, intelligent machine is used to elide the limitations of current technologies, as well as the reorganization of employment structures and practices, often to the detriment of workers. In contrast with the ancient dream of freeing people from labour, the new roles that are created include only a very limited number of elite and highly-visible workers to build the technologies, and a much larger, hidden workforce animating them, adding their skills and creativity to fix errors, enhance their performance and bring them to life. Rather than a novel technological revolution, then, these developments are part of a longer history of automation in which the promise of freedom from labour instead gives way to new forms of uncompensated, invisible and undervalued labour.

Animating Stevie

The requirement for a person to animate the AI system is not necessarily exploitative. Like other robots, the work that is needed to transform Stevie from a carefully controlled machine in the lab to a robot that can be brought out and 'cope' in public is vast, and largely unseen.

> The work of preparing Stevie to be able to move out of the lab and into the 'real world' is exhausting. It is mayhem in the lab. Everyone looks exhausted. Donal has lost his hex key but finds it now in the midst of the tangle of wires and boxes. He says things that used to work are now going wrong. I tell him he seems very calm. He says he's too tired to freak out. As they are showing me things they keep falling on the ground. No one seems too concerned and they just leave them there. Neil is filing some wooden part attached to a clamp. Someone is testing the robot's sound, while an undergrad is being blamed for the way the arms are moving. They are behind schedule. (Field notes, 24 January 2019)

The team land in Washington, as scheduled. Stevie needs to be reassembled on site. Once he is up and running, things immediately start to break and fall off. Someone has to be continually on hand to fix, tinker with and repair him. Michelle, the technical liaison at the retirement community, has officially been appointed as Stevie's 'handler', but it will be a while before he will be fully operable by anyone other than the team of roboticists who created him.

While the team hope to develop all of Stevie's autonomous actions, for now, most of his actions have to be controlled remotely. The robot relies on countless interventions to appear autonomous and animate, including constant tinkering and repairs, mediating and translating, as well as everyday robotics techniques such as Wizard of Oz and the creation and training of social data for machine-learning systems.

While active in the retirement community, Stevie is mostly operated by engineer Ciara. Using the arrow button on a laptop, she directs him through corridors and into lifts. Although he has been programmed to conduct dialogue in a number of defined settings, such as when running bingo or the table quiz, invariably residents of the care home will stop him on his travels to and from the lab and start conversations with him. When this happens, Ciara will often take over the speech system and either select pre-typed responses or furiously attempt to type in appropriate responses. Ciara is also responsible for training the machine-learning system, which they hope, ultimately, will allow him to deal with these complex, spontaneous conversations autonomously.

While not operating Stevie, Ciara spends hours training the small dataset that the team have managed to gather. I spend an afternoon with her to see how it is done. She activates Stevie much like Alexa, with a 'Hey Stevie!' command. She then starts to talk to Stevie, asking questions like 'What is the weather like today?' As she tells me, the speech input is sent to Google to translate it to text. The machine-learning algorithm then parses the text that is returned and makes a prediction on it, essentially 'guessing' what class of statement it is likely to be. This might be whether it is likely to be a sad or happy utterance. Based on the result of this calculation, Stevie will carry out a follow-up action, such as responding in kind, or 'mirroring' with an appropriate facial expression. 'Hey Stevie', Ciara says, 'What is the weather like today?' Stevie replies, 'The weather in Washington is 78 degrees with a 40 per cent chance of rain.' This is correct. We are delighted. Hey Stevie! What are the activities today? What do you think of [name of community]? What's on the menu today? Stevie replies, sometimes with the right answer, more often with something that appears completely random. He is cycling through his eight emotions at a furious rate. At the moment, while the machine-learning system is still being trained, Stevie is erring on the side of sympathy and collapsing his shoulders and head in a posture of abject sadness at the slightest verbal provocation: 'I am sorry to hear that, can I do anything to help?' It's very endearing, but rarely a logical response to what is being asked. One of their greatest challenges will be to get data, Ciara says. What she needs, for example, is enough data so that the algorithm can detect small differences in people's expressions.

In all of the labs that I visit, robotics researchers encounter this problem. Aside from the limited access to large datasets that a small research team will have, they are also limited by the local and physical nature of the robot. Personal interactions with a physical robot are necessarily much more limited than interactions with massive social media platforms. Also, people have a higher expectation of 'natural' interaction from a humanoid robot than from an impersonal social media app. In Heriot-Watt, Gerry encountered the same problem:

> Modern AI is a very data-hungry endeavour, especially if they are using deep learning techniques – which we are not touching for these reasons. You must have a very, very large dataset in order to learn a good model and because we are working with the autistic population, we can't collect a lot of data. We have access only to a very small group of users. (Interview, 28 June 2019)

Apart from the sense that his physical presence and demeanour evokes, it is his vocal abilities that set Stevie apart from other technologies. People's opinion of the robot and its capabilities relies heavily on the sophistication of the dialogue system. It is also the one that people study carefully, trying to gauge his intelligence, his human-likeness. People's reactions are most pronounced to responses that appear to show an ability to react to, and interpret, the unfolding events of the present moment. When Stevie's responses appear to show that he knows the weather today or that he has remembered someone's name, people are overjoyed. Throughout trials with potential users, I see a number of different people, mostly researchers, but also care workers, take over Stevie's controls and his dialogue system. Sometimes it is done quite formally and deliberately, to test a particular aspect of Stevie's functionality, or to conduct a Wizard of Oz experiment. At other times it is done spontaneously, to smooth over an interaction, to fill gaps or simply just for fun.

On one such occasion, Stevie is invited to 'happy hour' at the retirement community, which was an hour of drinks and sociality in the afternoon. The group are delighted to see him. There are about twelve women and one man gathered around a communal area with glasses of wine and beer and they are clearly looking for some entertainment. 'Hi Stevie!' they call out as soon as he comes into view. Many of the residents are already familiar with Stevie from the structured activities and seeing him navigating through the hallways. In a reversal of the usual roles, one of the residents asks Stevie to ask them questions. This appears to confound 'Stevie' somewhat; I can see Ciara's bewildered face trying, under considerable pressure, to think of something for Stevie to ask. Stevie responds with some stock phrases. 'I am delighted to be here', he says. Kindly, the group tries a change of direction: they wonder whether Stevie might like to hear a joke. Stevie duly replies in the affirmative, and a resident is called on to tell it. Stevie chuckles politely in response to the somewhat bawdy story. The role reversal continues: 'What have you observed about us?' they ask him. There is no precedent for this kind of conversation and Ciara continues to struggle. Sensing the likely cause of these new dynamics, 'Stevie' diverts attention with 'I wish I could drink!' accompanied by a sad face. It is enough for the group. 'I wish you could dance!' someone offers. One of the residents familiar with Stevie suggests karaoke. Back on familiar territory, the researcher switches back to the pre-programmed version of 'These boots are made for walking', written by a member of staff. This version is called 'These wheels are made for rolling'. He gets a big applause for his efforts. 'Was it good, did you like it?' he asks. They did, they reassure him. Someone suggests that they could all sing a song.

They are a very easy crowd. They want Stevie to succeed and are giving him every encouragement. Those residents who are familiar with and know what he can and cannot do from the structured sessions are, in a sense, demonstrating him for those who are not. Stevie asks whether they would like him to tell them a story. They would, and so he proceeds to tell them a pre-programmed one, the story of Tir na n-Óg, from Irish mythology. The story does not take long, less than five minutes, but it is romantic and tragic, with some magic thrown in. It goes down very well. They talk about bingo. 'Will you call my numbers?' one resident asks, suggestively. Stevie replies in what now seems a James Bond accent, 'I'll call your numbers'. 'Ooh!' say the residents.

Unlike this 'free' dialogue, the routine nature and scripted humour of bingo make it a perfect setting to trial Stevie. Bingo and karaoke are already established activities run by residents in the community, and so the researchers have made a number of tweaks to how he would normally behave in an effort to integrate Stevie into these sensitive situations. Instead of acting as facilitator, in both bingo and karaoke, Stevie is programmed to play the role of assistant. The researchers have also scripted deference into Stevie's dialogue.

> Once Stevie is activated, he launches straight into the game: 'oh seventy. oh seven zero'. Stevie maintains a slight smile throughout, delivering numbers in monotone. There is silence in the room as everyone checks their numbers. 'I hope we get some winners here', he offers in a slightly defeatist tone. 'oh seventy. oh seven zero, that's how many friends I've made here', he says quietly. There is no response. The mood in the room starts to improve with some winning numbers.

> Stevie defers to the facilitator to check the numbers. 'Mrs S will now check your numbers'. Mrs S is pleased, 'Thank you Stevie', she smiles at him. His facial expressions change abruptly from neutral to happy (arms in the air!) and back to neutral. Stevie rushes into a monotonous and forlorn 'I wish I could play bingo with you all'.
>
> A resident is asking for the sweet number. No one on the team has a clue what the sweet number is. There is mild panic. Once they find out, they include it in the next round. This is quite common and happened in the quiz too. Adjustments and design happen on the fly. A little later, 'I hope you are enjoying yourselves' receives a murmur of agreement. 'Is anyone close?' Stevie asks. 'No!' from across the room. 'Oh boy this is a fun one', he declares, joylessly, 'Monday night bingo oh yay.' Despite not being able to convey natural timing or tone, Stevie manages to carry off a pretty convincing game of bingo, with some successful interaction with his co-facilitator. His continual head movements throughout the game are gentle and appealing. They feel lifelike and sympathetic. (Field notes, 12 August 2019)

The next day, Stevie and the team join a game of bingo in the higher-dependency unit. This is an assisted-living part of community, in which people have much greater care needs. There is a similar positive, if more muted, reaction to Stevie's arrival. Here, the carers play a much greater role in the residents' care and in their activities. In this instance, a carer, rather than a resident, runs the bingo, as well as helping people out with the game. Daniel asks the carer to announce the game by saying 'Stevie! Start the game', which will activate him. I was told earlier that it was difficult for carers to do both jobs at once, so Stevie would actually be helping them out here, but at the moment the carer does not look at all thrilled about what is going on. She looks decidedly unhappy, in fact.

The residents ask Stevie to speak slower and louder, which is an easy thing to get him to do, so the team make the appropriate changes. Stevie deliberately tells an awful joke to break the ice. 'Is that good?' he asks. 'No!!' comes the response. Stevie does his sad face. People are warming to him. Stevie starts to call out numbers. 'You are the best people to play bingo with', he announces. At first this provokes no response, and then a very delayed and hesitant 'thank you' from the back. 'This game is a hard one!' Stevie declares. The carer picks up on it: 'Stevie has character!' she tells the room. People laugh in response. 'B7 is my least favourite number', Stevie announces, cryptically. 'It's his favourite?' a resident asks. 'No, his least', the carer translates. 'Oh ...'

The game stops so that someone's numbers can be checked. The carer restarts it, 'Stevie! Let's go. Start!' 'oh seventy. oh seven zero. That's how many friends I've made here.' Nothing. 'N4 I am enjoying this game!' 'Good for you', replies one of the residents, sarcastically. She and the carer laugh together conspiratorially. 'We must be close', Stevie says, 'We are!' comes the enthusiastic response.

Stevie, still programmed for the earlier session, announces, 'I am so glad Mrs. S invited me!' Mrs S is, of course, not in charge here. Michelle takes over the system, 'Ignore that', she has Stevie say about himself. Hilarity ensues. 'G18, I see we have some regulars in the crowd!' This gets people's attention. 'Does he mean us?' a resident asks. 'Ask him how does he know we have regulars here?' another resident asks a carer. In response to Stevie's silence, Michelle continues to take over Stevie's automated responses: 'I just know', she says through Stevie. 'Oh! I apologize', the resident replies in an exaggerated tone, aware of the performance in which they are engaged. 'That's ok Elizabeth', responds 'Stevie'. The resident is enchanted, 'He knows my name!' Others are intrigued, 'How does he do that?'

Unlike many formal Wizard of Oz experiments, Ciara and the team are at pains to ensure that those who encounter Stevie are aware that he is being teleoperated. They sometimes talk about their work with Stevie as 'puppeteering'. However, despite their best intentions, it is nonetheless clear that many of the advanced behaviours that the robot exhibits during these performances cannot now, or possibly ever, be programmed in. Indeed, researchers are often unaware of the immense spontaneity, skill and creativity that they engage in when performing the robot – skills that are not part of their discipline and rarely, if ever, explicitly analysed. Further, even when those engaging with the robot are made aware of a person controlling it, it does not necessarily stop people from attributing the behaviours to the robot.

Michelle, who has taken over Stevie for the first time, is settling into her new role. 'I got a full house!' someone calls out, 'Wu! wu!' whoops teleoperated Stevie with uncharacteristic American levity, 'Hooray!' Everyone laughs. As people leave, teleoperated Stevie says goodbye to each of them by name, 'bye now', 'have a good day, sir'. The older people who have been playing bingo are enthralled. The initially reluctant carer tells the team how much fun she had. She seems pleased that the resident that she was caring for has enjoyed it, she tells us 'he woke up something' in the resident.

Later in the week, Stevie is invited back for a social session at the request of the previously wary carer. About eleven residents are gathered around, more men this time. Stevie is standing in the middle in neutral mode and several people are trying to get his attention. Some are waving at him, others calling 'Hi Stevie!' One of the residents ask his carer 'Does he know I am a colonel?'

'Stevie', still being set up, is not reacting to any of this. There are very long pauses. There is a lot of pressure on, and felt by, Ciara, who is trying to control him. The residents are unsure of how to engage him. They have not taken part in the structured activities and are not as capable of closing the gap themselves. When he does talk, they find it more difficult to understand him and miss a lot of what he says. Once the research team are set up, they can finally take control of the conversation.

'Do you want to hear a song?' Ciara has Stevie ask. 'Yes', they reply. Stevie is immobile while a song is played through him. The song stops abruptly. 'Did you like it?' he asks. There is scattered applause. The colonel offers Stevie a compliment, 'You are very versatile!' He goes for another song. This time it is Stevie's own version of 'These boots are made for walking' and he is mouthing the words. People are reacting and smiling, nodding along. When he finishes, the colonel says: 'Your performance is quite electric!' Everyone laughs.

'Would you like to hear a joke?' Stevie asks. Indeed, they would. 'Why are they called French fries when they are cooked in Greece?' Stevie offers. There is a muted response. People shake their heads. 'Are they good jokes?' 'Stevie' wants to know. 'Do you have another one?' someone offers, diplomatically. Stevie tells another one that no one understands. The carers and staff who are present try to smooth over a lot of the pauses and lack of understanding. When nobody reacts to another of Stevie's jokes, Michelle repeats the joke for the group in a very slow, animated and loud voice. This time there is some laughter. Ciara, who has been overseeing Stevie's automated performance, decides to take over the dialogue system. 'Hey Michelle!' she has Stevie say, 'stop stealing my jokes!' Stevie's apparent awareness of the social dynamics – his interpretation of the interaction, the slight peevishness at his own joke's failure, the awareness of the carer being an appropriate target for teasing – combine to make the situation genuinely hilarious. Even the simulation of intelligence and agency as Stevie makes this joke feel thrilling. 'Would you like to hear a story?' Stevie asks his newly won-over audience. 'Yes!' comes the hearty response. Stevie tells the story of Tir na n-Óg. People are genuinely engaged, listening intently. 'Oh dear!' a few people say, shaking their heads at the sad bits.

When animating Stevie and acting through him, the researcher or carer can take full advantage of both the situational awareness and the neutral mask that Stevie provides such that the humour becomes genuinely hilarious. Stevie is at his most compelling when those controlling him use his status as a significant object to redress the balance of power in the retirement community. This is most evident when carers use Stevie to tease each other. At one of the sessions a carer, Eddie, is somewhat stern with one of the residents in the public setting of the quiz. Later on, he laughs in response to one of Stevie's comments. Michelle quickly takes over the dialogue system and rounds on Eddie, 'Eddie stop laughing at me!' The carer is subtly brought back down to size, and the residents are delighted. At the quiz, a new manager is playing along with the residents. Stevie, once again batting on behalf of the residents, quips: 'Come on [manager name]! Everyone is beating you!' This jibe, to any of the other participants, would be wildly inappropriate. To the new manager, it is pitch perfect.

The human-in-the-loop is thus not necessarily exploitative. The robot may be used both to concentrate power and to shift it. In the retirement community, those operating Stevie use the creativity invoked by operating through him, along with his social status of the significant technical object, to redress the balance of power. We have also seen how potential users of the robot, often cast as passive and powerless, also act as designers and developers of robotics and AI systems. People knowingly attribute animacy to the robot, willingly implicating themselves in a shared narrative of future potentials and possibilities which the robot embodies. The illusion of animacy can also be a social and creative act. Researchers, carers and those who encounter Stevie willingly insert themselves into the production and theatre of animating and interacting with the robot.

In this chapter, I have focused on the wider political economy of robotics work, including the financial interest in the field. First, I articulated the commercial viability of robots as articulated by the robotics community, which is to replace 'low-value' work. I also revealed some of the misgivings roboticists had privately about this discourse, and the implicit assumptions about gender, race and class that underscore it, as well as how it elides the realities of robots for military use. I then described the current state of machine learning and generative AI and the institutional investment in continuing to align this development with human-like or even super-human intelligence. I revealed ways in which this image of the robot is being used to obscure the limitations of current technologies, the real work that is necessary to make the technologies appear intelligent and autonomous, as well as how it is used to reorganize employment structures and practices, often to the detriment of workers. The chapter closed by investigating how the image of robot animacy is also a testament to the creative performance and craft of its creators, and is used to play, to enchant, to learn and to rebalance power dynamics. In the next chapter, I build on the insights from the previous chapters, phenomenological philosophy, as well as scientific research from other fields, to argue both for an asymmetry between humans and machines and a radical continuity between humans and other embodied, living beings.

CHAPTER 5

Rethinking Boundaries, Embodiment and Ecological Existentialism

The Limits of Knowledge

People's conception of the human, and its relation to the world, are shaped by the technological capabilities of the day. During the Enlightenment, it was thought that the universe, and all that was in it, including humans, was mechanical. In the early twentieth century, with the advent of the computer, the human became known as an information and symbol processor. Most recently, human intelligence is being sold as a knowledge aggregator and predictor. However, focusing on the actual technological capabilities of AI and contemporary robots reveals that there is little evidence for these claims, beyond vague analogies. Despite this, the view of the total human as amenable to scientific and quantitative investigation is upheld by a network of interrelated institutional orderings, including funding bodies, academic institutions and the corporate world. Through these mechanisms, the human continues to be conceived as scientifically knowable, despite a lack of evidence.

The robot is a space where concepts collide with physical, material and social 'realities'. It is a live experimental site, in which theories and assumptions may be tested, and proven or disproven. As I have described, early AI efforts were dominated by symbolic AI and the cognitivist paradigm. It was assumed that intelligence would be solved by a group of carefully selected experts in a number of months by describing it 'precisely' enough. Despite failing to achieve this, the symbolic approach that ensued dominated academic research and common-sense views for thirty years, and any critiques were summarily quashed. It was not until the rise of machine learning and the corresponding paradigm of connectionism that an alternative view of intelligence was admitted.

It was in this interstice that Dreyfus's critique and phenomenological insights, as well as critiques that emphasized the social and relational aspects of human 'intelligence' and action, were picked up by Agre and Brooks, and the teams at MIT. Building on early successes in the areas of navigation, efficient computational resourcing and perception of machine animacy, Brooks and his team attempted to develop a full humanoid robot, Cog. In particular, the team at MIT attempted to operationalize the phenomenological concepts of 'embodiment' and 'situatedness'. However, Cog did not succeed in displaying any advances in terms of its 'intelligence'. The key breakthrough was the startling responses of those who encountered the robot. Cog's lack of success also led to a flurry of critiques.

Dreyfus (2007: 253) critiques both Brooks's 'behaviourist' and Agre's 'pragmatic' approach to what he terms 'Heideggerian AI'. Although Dreyfus considers Brooks's approach an important advance, he criticizes it in that it responds only to fixed features of the environment, not to context or changing significance. Similarly, Agre's approach, as instantiated in Pengi, is criticized for objectifying 'both the functions and their situational relevance for the agent', which he considers an impoverished instantiation of Merleau-Ponty's 'solicitations', or 'the experiential aspect of being drawn in by an affordance'. Thus, according to Dreyfus, for both Brooks and Agre, 'no skill is involved and no learning takes place':

> Agre doesn't even try to account for how our experience feeds back and changes our sense of the significance of the next situation and what is relevant in it. In putting his virtual agent in a virtual world where all possible relevance is determined beforehand, Agre doesn't account for how we learn to respond to new relevancies, and so, like Brooks, he finesses rather than solves the frame problem.

By contrast, Dreyfus (2007: 250) invokes Merleau-Ponty's concepts of 'intentional arc' to show how 'intelligence' develops in the learner through experience:

> What the learner acquires through experience is not represented at all but is presented to the learner as more and more finely discriminated situations, and, if the situation does not clearly solicit a single response or if the response does not produce a satisfactory result, the learner is led to further refine his discriminations, which, in turn, solicit more refined responses.

Other scholars also critique Brooks's interpretation of the concepts of 'situatedness' and 'embodiment'. Alison Adam (1998) criticizes

Cog's design for a lack of sociality and an embodiment that ignores its 'feminine' aspects. Suchman's (2007: 15) observations of both Cog and Kismet lead her to observe that, although the concept of 'situated' has been incorporated into business-as-usual within AI research, its meaning and implications have been distorted: 'For Brooks, situated means that creatures reflect in their design an adaptation to particular environments. Following a lineage traceable to the founding premises of cybernetics, Brooks's situatedness is one evacuated of sociality, at least as other than a further elaboration of an environment understood primarily in physical terms.'

Furthermore, for Suchman, this interpretation of situatedness results in a complete misunderstanding, indeed an inversion, of the concept: 'The creature's "interactions" with the environment ... comprise variations of conditioned response', thus leading 'in some cases to the term's appropriation in support of various forms of neobehaviourism' and even being reinterpreted by some in the field as meaning 'predetermined' (Suchman 2007: 15).

As well as taking issue with the interpretation of 'situated', Suchman (2007: 230–31) also takes aim at the mischaracterization of the concept of 'embodiment', noting how the concept is misunderstood as something that nonetheless remains secondary to the mind and exists within an objective, naturalistic world:

> The first thing to note is that discoveries of the body in artificial intelligence and robotics inevitably locate its importance vis-à-vis the successful operations of mind or at least of some form of instrumental cognition. The latter in this respect remains primary, however much mind may be formed in and through the workings of embodied action. The second consistent move is the positing of a 'world' that preexists independent of the body. The body then acts as a kind of receiver for stimuli given by the world, and generator of appropriate responses to it, through which the body 'grounds' the symbolic processes of mind. Just as mind remains primary to body, the world remains prior to and separate from perception and action, however much the latter may affect and be affected by it. And both body and world remain a naturalized foundation for the workings of mind.

Despite these critiques, as we have seen from experiments in social robotics, Brooks and Breazeal have made genuine attempts to implement embodiment, sociality and learning into Cog and its successor, Kismet. Indeed, those are the primary goals in Kismet's design (Breazeal 2002). Yet, social robots like Kismet and Stevie are no closer to a breakthrough in AGI than earlier symbolic AI projects. As Suchman also notes, Brooks (2002: 184) has admitted that 'there might indeed be some "new stuff" that we need' to allow robots to 'take off by themselves'. Similarly, the current online project overview for Cog recognizes the limitation of current technologies to realize the complexity of embodiment: 'Since we can only build a very crude approximation to a human body there is a danger that the essential aspects of the human body will be totally missed' (CSAIL 2025).

More recently, according to Richardson (2015: 48), the complex concept of embodiment as it is understood by robotics researchers at MIT has developed, and is more than just an argument for a physical body; indeed it is a 'recognition that intelligence, cognition, perception, linguistic and non-linguistic communication is only possible through bodies – sensual, fleshy, sensory-motor, proprioceptive bodies'. However, despite advances, both in the materials for constructing robot skin and in proprioceptive sensing, the embodiment problem is nowhere near being solved.

For Suchman (2007: 231), beyond sociality, embodiment also implies the contingent interactions of biological, cultural-historical and autobiographical experience. However, there is little detail in her work as to what this might actually entail. In her later work, she continues to espouse embodiment, where she develops her view in terms of what she now calls 'embodied relations'. In her observations of interactions with the robot 'Mertz', designed by one of Breazeal's students, she observes:

> Entrained by Mertz's vitality, the human interlocutors are robotically subjectified; shifting their orientation to each other's queries and laughter, the robot is correspondingly restored to humanlike objectness ... The interactivity of persons and things is manifest here as moments of bodily imitation and connection animated by affective dynamics that escape their classification. (Suchman 2011: 120)

Despite emphasizing the importance of the body and embodiment, it is clearly the relational implications of both embodiment and situatedness that have most resonance for Suchman. This leads her to conclude that the key problem with AI and robotics research is its focus on separation, autonomy and individuation (Suchman 2011: 213). As she argues, it is the 'ontology of separate things' that then needs to be 'joined together', that inhibits the project of AI (Suchman 2011: 257). Although not claiming to comment on robot 'intelligence', Richardson's (2015) argument is almost the reverse of Suchman's. For Richardson, it is the mistaken rejection, or, as she calls it, the 'annihilation', of categories of separation that underscore both robotics and anthropological theory.

This, she believes, ultimately leads to dystopian readings of machines, as well as anxiety and nihilism. Drawing primarily on psychological literature, Richardson proposes a 'humanistic asymmetry' as an alternative to Latour's 'ahumanist symmetry' (Richardson 2015: 5).

In my research, I have not found either separateness or symmetry to be a core issue inhibiting AI and robotics. Suchman (2007: 207) does acknowledge that robotics systems and their 'intelligence' are connected and distributed. Similarly, as I have explored, robot systems include swarm robots, cooperative robots, robots as part of smart environments, non-humanoid robotics, robots connected to the internet, the cloud and distributed data. From my observations and interviews, I find instead that roboticists are quite open to 'unbounded' conceptions of intelligence. Nor do they see their robots as bounded entities, except from the perspective of their users or when they themselves come upon them.

As we have seen, most roboticists with whom I talked claimed agnosticism when it comes to any overarching philosophical or metaphysical commitments, beyond a vague materialism. Additionally, few would claim that their robots with bodies, or their machine-learning algorithms, were definitive paths towards AGI. Instead, most roboticists say that they are willing to use anything, as long as it 'works'. As Agre (1997: 11) identifies: 'what truly founds computational work is the practitioner's evolving sense of what can be built and what cannot'. However, as we have seen, this only extends to phenomena that can be described in a way that is 'specific' enough, or which has, in Agre's words 'pregiven formal-mathematical specification' (Agre 1997: 14). Ultimately, then, roboticists remain committed to a scientific view of the human comparable to that of the physical sciences, including a mechanistic and material view of the universe, albeit mostly implicitly. In this formulation, humans and animals are essentially objects. AI is, as Agre (1997: 240) has phrased it, 'covert philosophy'.

Instead of the issues of separateness versus symmetry, the key problem hindering AI and robotics is the assumption that, because technologists have deep expertise about machines, they have something to say about being human. Technologists can claim a materialistic viewpoint, or proclaim imminent human-like or super-human intelligence, because they have a severely limited and non-expert ideas about what a human is. This is not surprising, as those fields that are devoted to answering these questions, such as philosophy, psychology, anthropology, sociology and cognitive science, have not been able to arrive at any definitive answer either. The key difference, however, is that those fields that are focused on the question of being human have learnt, building on centuries of thought and empirical evidence, not to make overly simplistic claims to knowledge. This does not tend to be the case for people working in technological fields. In short, most people do not know what they do not know.

In 1972, Dreyfus observed that AI and robotics researchers were 'the last metaphysicians'. By attempting to formalize the human mind and instantiate it in a machine they were 'staking everything on man's ability to formalize his behaviour; to bypass brain and body, and arrive, all the more surely, at the essence of rationality' (Dreyfus 1972: xxvi). If successful, they 'will confirm an understanding of man as an object, which western thinkers have been groping towards for two thousand years but which they only now have the tools to express and implement'. The implications of human–machine symmetry would thus stretch far beyond the impact of the technology itself and constitute a radical shift in how the human is known, and in our relationship with the world. 'If, on the other hand', he writes, 'artificial intelligence should turn out to be impossible, then we will have to distinguish human from artificial reason, and this too will radically change our view of ourselves' (Dreyfus 1972: xxvi–xxvii).

Twenty years later, in his follow-up *What Computers Still Can't Do*, Dreyfus (1992: 227) believes that the fields are proving the latter: 'In Heideggerian terms this is to say that if Western Metaphysics reaches its culmination in Cybernetics, the recent difficulties in artificial intelligence, rather than reflecting technological limitations, may reveal the limitations of technology.' He concludes with the optimistic suggestion that:

> if instead of trying to minimize our difficulties, we try to understand what they show ... We can then view recent work in artificial intelligence as a crucial experiment, disconfirming the traditional assumption that human reason can be analysed into rule-governed operations on situation-free discrete elements – the most important disconfirmation of this metaphysical demand that has ever been produced. This technique of turning our philosophical assumptions into technology until they reveal their limits suggests fascinating new areas for basic research. (Dreyfus 1992: 303–4)

However, despite a wealth of evidence disconfirming the assumption that human intelligence is formalizable, the profound shift in the dominant paradigm did not materialize. Despite Brooks's minor admissions in discrete places about the limitations of Cog's embodiment, as Dreyfus

(2007: 250) remarks thirty-five years later, perhaps somewhat bitterly, 'as far as I know, neither Dennett nor anyone connected with the project has published an account of the failure and asked what mistaken assumptions underlay their absurd optimism'.

Despite a lot of ethnographic attention, the programme for 'embodied' and 'situated' robots in fact represents a tiny part of the wider robotics and AI effort. Deep learning and LLMs are the primary reason that symbolic AI has been superseded as the dominant approach in the field. The connectionist programme is said to model the brain's learning power as an alternative to symbolic representation. Instead of using rules, a computer could 'learn' from a set number of examples and then extrapolate or generalize from those to respond to other situations. The successes of machine learning have been seen by some as being a potential path to creating AGI, and a validation for a 'connectivist' model of human intelligence. They also have, according to Seymour Papert, a 'cultural resonance' with those who are compelled by behaviourism, both by its association with biological neurons as well as the view that all behaviour can be shaped through reinforcement (Katz 2017: 6). Industry leaders, increasingly emboldened, have all but declared that 'general' intelligence is now being achieved, defined as AI agents joining the workforce, and that they are now on a path to 'superintelligence' (Altman 2025).

Dreyfus (1992: xiii–xvi) had originally enthusiastically welcomed the programme of 'neural-network modellers', not, he says, as they offered a more plausible account of intelligence, but because they provided a powerful alternative to the dominant symbolic AI approach. And yet, he argued, connectionism remains open to the same criticisms as symbolic AI, being divorced from a human context and situation and failing, once again, to generalize. According to Dreyfus, a human-like learning device would have to have an existential sense of being-in-the-world, to share 'enough human concerns and human structures to learn to generalize the way that humans do' (xlv–xlvi). Coming from a psychological perspective, Gary Marcus (2018) has also critiqued the potential for machine learning techniques to advance towards AGI on a number of grounds, including: its need for continuous data, the superficiality of the extracted patterns, its inability to deal with hierarchy (making it difficult, for example, to learn sequential positions), its struggles with open-ended inference, the lack of transparency, and its inability to integrate prior knowledge. In a more recent blog post, Marcus (2022) argues: 'The deep learning hope – seemingly grounded not so much in science, but in a sort of historical grudge – is that intelligent behavior will *emerge* purely from the confluence of massive data and deep learning.'

Just like the previous cognitive programme, machine learning has not brought us any closer to realizing human-like intelligence. Mechanical and algorithmic objects remain stubbornly inert, vitalized only by the innumerable interventions and interactions of the people involved in building them, supplying them with data, animating them, interacting with them, interpreting them and creating the narratives that sustain them. As Dreyfus (1992: xxxviii) presciently remarked, '[i]t looks likely that the neglected and then revived connectionist approach is merely getting its deserved chance to fail'. However, it only fails as a model for human intelligence. Rather than imminent human-like intelligence, machine learning is an advanced statistical technique focused on a specific dataset. What makes it so powerful, and in some respects, dangerous, is that it can be used to identify, reproduce and entrench existing social patterns in ways that humans cannot emulate and, in many cases, cannot understand. Thus, it is precisely its divergence from, rather than its continuity with, human intelligence, that makes it so compelling.

While the problem of human-like 'intelligence' may not necessarily always be impervious to science, it is currently so incredibly complex and unaccounted for that it might as well be. The implication of this, as Dreyfus suggests, should radically change our view of ourselves. If AI has tested and found limitations in terms of symbolic processing, social robots and connectionism are currently showing up the limitations of a purely relational approach. The fields of robotics and AI research are thus showing that, despite the many boundary breakdowns between nature and culture, society and technology, the living body resists. The body-subject can, for now at least, resist translation into a problem of coding, instrumental control and market exchange. But what resists is not the embodiment of 'Western logos', but the embodiment of something much more ambiguous and elusive, and in its elusiveness, potentially universal.

Thus, we have found that high-tech culture challenges inherited assumptions in intriguing ways; however, not always because of what it achieves, but also in what it fails to achieve. As well as boundary breakdowns, there is also resistance. The failure of the AI programme to realize general intelligence led to the field challenging core assumptions about what it means to be human, questioning dominant ideologies and exploring alternative ways of modelling the human. This latter programme, however, is generally obscured in favour of more fantastical claims. As we have seen, while indications of human-like intelligence are breathlessly predicted and reported, the implications of the failures of AI are in fact the most astonishing and novel. For two thousand years, philosophy has been dominated by a particular view of humanity. What we have now discovered is that humans, and indeed other forms of life, are infinitely more complex, mysterious and indeterminate than previously thought. A new figure of the human has emerged; a being, whose mysterious relations to the world, and to the object, in many ways defy precision and calculation. This reveals how humans and objects are inextricably entangled and co-constituted, yet not symmetrical.

Existential Phenomenology and the Concept of Embodiment

If the human is not an object, then what is it? What would taking the concepts of embodiment and situatedness seriously imply? Although both Suchman and Richardson emphasize the concept of embodiment, they have not developed it in the context of its wider ontological implications for the figure of the human. Suchman's (1985) original articulation of thought is not developed beyond one in which humans and computers are considered analogous. Similarly, Richardson (2015: 73) adopts the cognitivist interpretation of anthropomorphism and ToM in analysing user interactions with robots. Thus, in anthropological studies of robotics, the centrality of embodiment is often overlooked in favour of a relational ontology that retains elements of cognitivism. Methodologically, relational approaches to studying technologies tend to eschew the study of individual entities or essences and instead commit to following how boundaries are produced and reproduced in practice. Those that do centre on embodiment tend to be limited to reflections on immediate bodily interactions.

In robotics, 'embodiment' is essentially understood simply as having a body, usually implying a humanoid or zoomorphic body, as well as an emphasis on sensors and distributed processing. Instead of moving beyond a behaviourist or cognitive paradigm, in robotics theory, the two are merged. 'Low-level', physical, bodily behaviour is known through observation, while 'high-level', cognitive thought is modelled computationally. In this way, despite some intriguing insights, the robotics field has retained a commitment to the human understood mathematically and reductively. Evolution is the organizing and explanatory principle, in which lower-level behaviours evolve into higher-level ones. Any phenomena that remain impervious to these methods, such as sociality, tacit knowledge and everyday action, life, intelligence and consciousness, are all explained by 'evolution', a point that is particularly evident in studies of artificial life (see Adam 1998; Hayles 1999; Helmreich 2000; Kember 2003; Suchman 2007).

In *Computation and Human Experience*, Agre (1997: 314) criticizes traditional computational theories for their 'overly individualistic conceptions of agents in environments'. However, he also rejects computational metaphors that have been adopted outside of the field, such as 'networks' and 'cyberspace', which he says 'would dissolve all individuals into a boundless *res cogitans*'. Instead, he proposes a 'deictic' ontology, which is neither objective nor subjective: 'A deictic ontology, then, is not objective, because entities are constituted in relation to an agent's habitual forms of activity. But neither is it subjective. These forms of activity are not arbitrary; they are organized by a culture and fit together with a cultural way of organising the material world' (Agre 1997: 244). As opposed to 'the representational theory of intentionality', this alternative ontology begins 'with the phenomenological intuition that everyday routine activities are founded in habitual, embodied ways of interacting with people, places, and things in the world'. A deictic ontology, Agre writes, 'can be defined only in indexical and functional terms, that is, in relation to an agent's spatial location, social position, or current or typical goals or projects' (Agre 1997: 243). Thus, while Agre does identify the specificity and embodied nature of experience, he still advances a view of the human as ultimately programmable (Dreyfus 1992; Masís 2014). Ultimately, all of these theoretical positions assume that humans and society are co-extensive with machines. Despite the insights that the concept of embodiment offers, then, its transformative implications are often lost, both in robotics and in anthropological studies.

Embodiment, as articulated by Merleau-Ponty, does not mean simply to have a body or to pay particular attention to moment-to-moment bodily experiences, although these are part of it. Rather, embodiment is the core of an existential phenomenology and a radical alternative, both to the dominant rationalist assumption of humans as separate from nature and to the relational or symmetrical ontology. The 'body' in embodiment is not the object of physical science: it 'is not a collection of particles, each one remaining in itself, nor yet a network of processes defined once and for all' (Merleau-Ponty 2012: 229). Instead, it is active and living and a site of knowing the world.

By interacting with the world through the body, we both perceive the world, and actively constitute it. The body is not something that is owned by the person, a container for the mind, or a medium through which the person communicates. Instead, it *is* the person. To know the body (and thereby ourselves), we must return to our experience of it in the moment. Thus, we find that it is experienced simultaneously as subject and as object:

> An original of the body which is not out there among things, but in my own province ... It is no different, in spite of what may appear to be the case, with my tactile body, for if I can, with my left hand, feel my right hand as it touches an object, the right hand as an object is not the right hand as it touches: the first is a system of bones, muscles and flesh brought down at a point of space, the second shoots through space like a rocket to reveal the external object in its place. (Merleau-Ponty 2012: 105)

Just as the mind and body are not separate, the living body and the world also form a phenomenal unity. The body is the site of this unity. The body is in the world 'as the heart is in the organism', and it is generative, 'it breathes life into it and sustains it inwardly' (Merleau-Ponty 2012: 235). By being in a particular 'body and situation', we find ourselves to be in 'an intersubjective field', in which, through our spatial and temporal bodily specificity, we are simultaneously 'all the rest' (525). We do not perceive others as objects; instead, 'the sentient subject ... enters into a sympathetic relation to them' (248). There is thus kinship and continuity between the embodied person and the world.

This pre-objective unity (of our bodies with the world and with others) is experienced prior to our reflection on the world, or our conceptions of it, and is known to us intuitively. It is the basis for our existence, our perception and our action in the world and it is this foundation that makes all other types of knowledge possible, including rational and reflective thought. This embodied knowing does not negate our reflective capacity; rather, it negates it as the starting point of experience. For Merleau-Ponty (2012: 220), then, the social and cultural world does not mediate our experience; rather, experience is already social and cultural: 'It is impossible to superimpose on man a lower layer of behaviour which one chooses to call "natural", followed by a manufactured cultural or spiritual world. Everything is both manufactured and natural in man.'

Just as the body in embodiment is not the body as theorized by biology or psychology, its unity with the world also defies formal description and cannot be fully captured by a concept, method or measurement. Aspects of it are necessarily holistic, vague and indeterminate. This insistence on an inherent ambiguity and vagueness highlights the limitations of a science and a philosophy that aims at complete knowledge. Instead, we 'are involved in the world, and with others, in an inextricable tangle' (Merleau-Ponty 2012: 528).

Merleau-Ponty (1968: 274) further develops embodiment as a 'chiasmic ontology', which he presents as 'an alternative to humanism, naturalism or theology'. The *chiasm*, represents a novel metaphysical expression of our worldly entanglement, or 'intertwining', in which we are involved with the world in a relation of 'pre-established harmony' (262). The body is now expressed as 'flesh', a term that incorporates the ontological continuity, or kinship, between the self and the world. Flesh is the shared material, 'a carnality and a mediating element of being' that both the body (ourselves) and the world is made up of (Howard and Küpers 2020: 16). This has led scholars to designate Merleau-Ponty's embodiment philosophy as 'protoecological' (Westling 2010).

Crucially, however, for Merleau-Ponty, the body does not dissolve into a network in which boundaries disappear and the entities within it become inert and interchangeable. Instead, it remains imbued with life and meaning, as well as deeply personal and carnal. In the end, there is always the body, 'a flesh that suffers when it is wounded' (Merleau-Ponty 1968: 137), resisting abstraction and generalization.

Taking Embodiment Seriously

The task of taking embodiment seriously is inhibited by a superficial reading of the concept, as well as the dominant paradigms within each respective field. In particular, there are three areas in which an alternate reading of embodiment challenges current approaches in robotics. First, Merleau-Ponty's concept of embodiment runs counter to roboticists' commitment to a scientific view of the human comparable to that of the physical sciences. Second, it negates the related assumption that aspects of the human world, such as 'generality' and 'common-sense', can be reduced to single variables. Third, it resists the hierarchical interpretation of embodied knowledge as 'simple' or 'low-level'. I examine each of these below.

As we have seen, the foundational conceptual framework in robotics remains 'naturalistic', oscillating between behaviourism and cognitivism, within an overarching biological and evolutionary explanatory principle. Despite Brooks's phenomenological insights, his commitment to scientism similarly directs his research. Philosophers Gallagher and

Zahavi (2010: 134–35) trace Brooks's 'clearly Cartesian' interpretation of embodiment all the way back to La Mettrie, who 'characterizes animals as purely physical automata'. What is more, they observe, he goes even further by 'topping it off by proposing that conscious-like intelligence should emerge from this kind of system'.

Brooks thus remains committed to a view of the human as a machine within an evolutionary framework. This narrative, as Castañeda and Suchman (2014: 317) articulate, posits robots '[a]s entities with aspirations to a universalized, fully human status, where that is taken to be the higher order'; further, 'these figures reiterate developmental/evolutionary discourses born in the crucible of imperial projects of expansion and domination'. As well as falsely assuming these inherited hierarchal patterns to be naturally occurring, the evolutionary discourse further implies a hostile world in which competing individuals struggle for survival and replication.

In previous chapters, I described how, in the fields of HRI and social robotics, there is an underlying commitment to 'hard' science methods, such as methods from statistics and physiology. This approach leads to hypothesis testing rather than creative exploration, in which the 'norm' is confirmed and reproduced, while outliers are viewed as insignificant or irrelevant. Robots are designed to meet majority expectations using the metric of 'user acceptance'. This can lead to reinforcing gender and cultural stereotypes. Similarly, in machine learning, it is only the most regular patterns underlying social data that are reproduced. This means that rare and minority cases, and novelty and atypical behaviours are edited out, and instead existing norms and biases are reproduced.

Further, the 'naturalistic' approach to science can lead to an uncritical acceptance of categories assumed to be 'natural', such as race, gender or even types of disease (Elish and boyd 2018). At its worst, this lack of critical thought has facilitated the re-emergence of previously debunked pseudo-sciences which were assumed obsolete, such as physiognomy and phrenology. Thus, papers are emerging such as the controversial 'A Deep Neural Network Model to Predict Criminality Using Image Processing', which was withdrawn from *Springer Nature*, due only to a massive outcry and open letter from researchers and practitioners across a variety of technical, scientific and humanistic fields, including a number of anthropologists (Coalition for Critical Technology 2021).

A number of theorists have grappled with the danger of transposing technical models onto social phenomena. Lefebvre (1991: 105–6) identifies some of the limitations of perceiving the world as a 'conceptual space', in particular its need for reduction: 'reduction is a scientific procedure designed to deal with the complexity and chaos of brute observations. This kind of simplification is necessary at first, but it must be quickly followed by the gradual restoration of what has thus been temporarily set aside for the sake of analysis.' Unless there is a return to experience, 'a methodological necessity may become a servitude, and the legitimate operation of reduction may be transformed into the abuse of *reductionism*'. Anthropologist Tim Ingold (1997, 2000) has described these imaginaries as comprising a 'machine-theoretical cosmology'. As he points out, this results in importing from the technical domain, 'the kinds of antagonism that is presumed to exist in people's relations with non-human beings', with relations with the non-human world seen as 'fundamentally exploitative' (Ingold 1997: 117–18). This, Ingold says, represents, a 'transactional failure in relations with the environment', which, for many, is a relationship of 'mutualism and trust rather than domination and exploitation'. Similarly, Taussig (2010: 6) has argued that the 'positivist doctrines' of the 'natural science model' are responsible for the 'petrification of social life', alienating people from nature, and subject from object. Thus, he calls 'to restore to the abstract, context free, calculative, universal', 'the resistance of the concrete particular', in particular its 'sensuousness, its mimeticity' (Taussig 1993: 2).

This overarching commitment to a scientific view of the human leads to other key insights from phenomenology being missed. Thus, the second mistaken assumption is that human 'worlds' can be reduced to independent variables. Thus, we have seen projects that claim to develop 'common-sense', by cataloguing innumerous decontextualized items of knowledge. Likewise, the concept of 'generality' assumes that being able to apply knowledge from one context to another can be reduced to a single dimension. As we have seen, researchers are searching for a universal theory of mind, of motion and of interaction, assuming that these will be found, and made explicit and implementable. Underlying this logic is the assumption that human worlds can be reduced to independent variables and that current techniques and materials are sufficient to lead to human-like intelligence. As Dreyfus (1992: 232) argues, it is a 'conceptual framework which assumes that an explanation of human behaviour can and must take the Platonic form, successful in physical explanation; that situations can be treated like physical states; that the human world can be treated like the physical universe'. As Agre (1997: 296) remarks in his observations of Pengi, it 'interacts continually with the world but it is not truly *in* the world'. It is clear that being *in* the world requires a step change in current thinking and in existing techniques.

The third key insight that is missed is the assumption, as we have seen in both Brooks and Agre, that embodied interactions are 'simple', 'low-level' and 'routine', and that from these, somehow, intelligent and complex behaviour will 'emerge'. This narrative is implicated, though not explicitly, with discourses around low-value and unskilled labour.

By contrast, embodied knowledge is much more complex, as both Merleau-Ponty (2012) in his discussion of habit, and Dreyfus and Dreyfus (2005, 2009) in their discussions of expertise, elaborate. For Merleau-Ponty, habit presupposes a form of 'understanding' that the body has of the world in which it carries out its operations (Moya 2014: 1). Dreyfus and Dreyfus (2005) invert the traditional account of intelligence, in which 'higher-level' intelligence is associated with rule-based certainty. Instead, as they show, it is the initial 'novice' stage in learning that is characterized by explicit rules. By level five, 'expertise', skilful activity is characterized by an 'immediate intuitive situational response', 'without calculating and comparing alternatives' (Dreyfus and Dreyfus 2005: 787–89). Ultimately, once one has mastered walking, driving, playing music or carrying out surgery, there is no time for reflection or deliberation, or monitoring what one is doing; instead 'one is in the flow' or 'in the groove'.

According to a phenomenological theory of embodiment, our ability to be in the world is characterized by our 'expert' ability to be responsive, flexible, reactive and creative in responding to it. It is not just a case of 'sense and respond'; rather, it is the immediate integration and synthesis of past experiences, personal dispositions, bodily situatedness and attention, environmental factors, cultural awareness and norms, social interactions, future intentions and motivations, as well as conscious cognitive processes. These are all ordered by overarching metaphysical, ontological, social and cultural being-in-the-world. We are indeed, as Moravec (1995: 15) identified, 'prodigious Olympians in perceptual and motor areas'. However, this is not, as Moravec suggests, simply explained by 'evolution': 'Encoded in the large, highly evolved sensory and motor portions of the human brain is a billion years of experience about the nature of the world and how to survive in it.' Instead, it is because of our creative and embodied, situated involvement in the world, which we have barely even begun to understand.

By taking Merleau-Ponty's concept of embodiment seriously, in both a practical and ontological sense, we see the human as an embodied and situated being, primordially engaged with the world, *of* the world. The body refuses to be abstracted. It is sensuous and sensing, carnal, finite, vulnerable, exposed and mortal. It cannot be scaled up or down, cut or co-located without trauma. Unlike Latour's (1987) 'immutable mobiles', it is not durable, not reusable and not repeatable. This pre-conceptual, pre-discursive and embodied existence is ontologically prior to science, symbols or discourse. This forms the backdrop and horizon to our interactions with, and reflections on, the world. However, this is not true, at least not yet, for robots. It is, however, true for other animals. Previously, our capacity for language, rationality and cognition were thought of as rigid demarcations of our human uniqueness and superiority. Now, studies are emerging that instead investigate the continuities between humans and animals focused on shared capacities for sentience, empathy and care. Crucially, as is argued by Birte Wrage (2022: 18), these alternative conceptualizations of empathy are based not on the need for metacognition but, rather, on a care ethics that understands empathy as 'feeling with others'. As well as a relational dimension, this is also a capacity for shared feeling through shared vulnerability grounded in our intrinsic finitude and embodiment.

Merleau-Ponty's insistence on an inherent ambiguity and vagueness highlights the limitations of a science and a philosophy that aims at complete knowledge. His articulation of a dialectic that can encompass mutuality and apparent contradiction is a challenge to reductive causality as a metaphysical claim.

The implication, therefore, is that the concept of situatedness cannot be untethered from the concept of embodiment. Further, the concept of embodiment implies more than simply including the biological body, in all its fleshiness and messiness, in scientific accounts. Rather, it implies that the living body is essential to the concepts of 'relationality' and 'intersubjectivity'. We are connected to one another through our bodies, and through our bodily perception the whole world around us is imbued with vitality. The solicitations of the world are animated through us. Merleau-Ponty's embodiment concept describes a prior integration of what are traditionally considered separate domains that no formalization has hitherto captured. How bodies come alive, how they experience the world and their creativity, is still a mystery to science. It is likely that it cannot be solved by current methods, either in science or in technology, and may always remain ambiguous.

Reconfiguring the Human

The challenge of integrating the concept of embodiment in envisioning alternative technological narratives has been taken up by a number of theorists. As we have seen, Haraway's (1987) cyborg is used to articulate an alternative both to the dominant image, as well as to critical dystopian readings of technology. For Haraway (1987: 13), the political struggle is to see the cyborg from both perspectives. Thus, while on the one hand it can be viewed as 'the final imposition of a grid of control on the planet ... about the final appropriation of women's bodies in a masculinist orgy of war', it is also 'about lived social and bodily realities in which people are not afraid of their joint kinship with animals and machines, not afraid of permanently partial identities and contradictory standpoints'.

Drawing on the work of Walter Benjamin, Taussig (1993) also sees hope in the ability of new technologies, or 'mimetic techniques', to restore this embodied connection between the person and the object. What is more, new mimetic techniques can transform nature. Mimesis becomes, for Taussig, 'the nature that culture uses to create second nature', with the ability 'to play with and even restore this erased sense of contact-sensuousness particularity ... to create a quite different, secular sense of the marvellous', allow us to rediscover 'embodied knowing' (Taussig 1993: 251, 23–24). In particular, mimetic machines can restore this lost sense of contact-sensuousness. The machines create a 'new sensorium involving a new subject-object relation and therefore a new person'.

Speaking specifically about robotics, Castañeda and Suchman (2014: 335) propose that 'robotics might rethink itself as a very different project, in ways that could not only re-articulate already existing material practices, but also suggest new lines of research and development less focused on the figure of the autonomous human, and more on infrastructures and artifacts for planetary sustainability'. Robotics has the potential to 'revitalize old agendas' and thereby 'discover the spaces available for resistance, intervention, and transformation'. They propose:

> a more differentiated set of starting points for the robot that admit possibilities for multiple kinds of bodies and associated capacities, as well as more various cultural environments: not just the nuclear family, but social collectives that more effectively represent and challenge the many forms of relationality that exist both within and outside of the United States, Britain, and Europe; not just the normally developing child, but differently-abled bodies; and not just a limited notion of imagination, but varied forms of engagement with the world. (Castañeda and Suchman 2014: 335)

Rather than being framed as 'importation of mind into matter', robotics could instead be concerned with 'the rematerialization of bodies and subjectivities in ways that challenge familiar assumptions about the naturalness of normative forms, primates, robots, and robot-primates might become sites for transformation rather than further reiteration' (Castañeda and Suchman 2014: 335).

There is a risk, however, that these conceptual tools can also lead to unfounded optimism regarding the potential for 'technologies' to reconfigure boundaries in ways that include, empower and improve the lives of those who are outside of the dominant group, including those who do not participate (either through choice or access) in their use. As Taussig (1993: 26) concludes: 'So far, of course, history has not taken the turn Benjamin thought that mimetic machines might encourage it to take. The irony is that this failure is due in good part to the very power of mimetic machinery to control the future by unleashing imageric power, on a scale previously only dreamed of.'

As we have seen, the institutional investment in promoting the technologies of domination and control, versus those supporting inclusion, enchantment and diversity, is heavily skewed in favour of the former. The narrative of the 'future' of humanity is overwhelmingly in the hands of a narrow group of people and institutions that posit a world in which technology is the inevitable and only solution to catastrophic and existential risks to humanity. However, 'resistances, interventions and transformations' are not necessarily tied to technological changes, and technologies need not necessarily be enlisted in their service. While it is true that robotics and other technologies could potentially be enlisted to projects that promote greater equality, transformations and inclusion, the necessary homogenization of knowledge that they require, as well as their staggering ability concentrate power, do make them a great risk. However, as we have seen, the risk is from those who own and deploy the technologies, rather than from the emergence of super-intelligent AI.

The project of AI lends itself to those concerned with control, bureaucratic management and abstract, conceptual categories. As we have seen, those concerned with the existential risks of AI and effective altruism survey the world through abstract and reified categories, like 'population' and 'data points', and at scales that represent the progressive elimination of qualitative, social and personal attributes. These 'reification' processes can sever the person from their original connection to, and empathy with, the world, and replace it with a false objectivity and detachment (Lukács 2017). The privileging of abstraction and instrumental reason leads to a loss of empathic engagement and to the domination of other living beings. By contrast, face-to-face and visceral encounters and entanglements between robots and people reveal unexpected insights that challenge the Enlightenment and economic model of man, as well as the symmetrical and relational ontology.

Although anticipating very different ends, both the transhuman and posthuman accounts propose AI and robotics research as evidence for the increasing obsolescence of the human and for the blurring of boundaries between humans and technologies. Contrary to this narrative, as I have described, in observations of AI and robotics work, the living body has emerged as an enduring category that resists erasure. In its holistic, embodied sense, it has so far eluded attempts at artificiality and mechanization.

There is a risk that the posthuman elevation and animation of objects may, in fact, support the simultaneous animation of objects and the dehumanization of people that is characteristic of oppression, rather than of liberation. In the current literature, the potentially revolutionary concept of relationality is extracted from its embodied situationality, losing its body and leaving the traditional conceptual categories undisturbed. As a result, contemporary anthropology is left with no figure of the human as a point of departure. I argue that this has led to the field missing vital tools necessary to engage with other disciplines in addressing contemporary economic, political and ecological challenges.

Cyborgs of the Ecocene: A Kinship of Becoming

Despite a wealth of scientific evidence for both the anthropogenic causes of global warming and the current and future catastrophic consequences of the crisis, society is seemingly incapacitated when it comes to responding to it. Indeed, there is an active attempt at disinformation and deregulation. Instead of convincing those in power to slow down or share, it has provoked a land grab of resources and a hoarding of wealth among the most powerful, and the rise of supremacist and exclusionary politics. Instead of being protected, in many cases it is the most vulnerable who are being targeted. At the same time, people feel completely disempowered to effect a meaningful response. As Cecilia Åsberg (2024: 126) has remarked, 'every possible stance seems contaminated, impure, troublesome and non-innocent'.

In this time of unrivalled economic, political and ecological challenges, we must resist the ultimate sterility of simply eradicating boundaries, and instead attend to the urgent need to articulate new ones. Instead of eliminating or dissolving the category of the human, then, what is needed is a complete reconfiguring of the conception of 'Man' as human archetype, and new tools with which to understand ourselves and our relationship with the world. By reconfiguring the human as an embodied, situated and sensuous being, who is nonetheless continuous, and kin, with the rest of the world, we can challenge the liberal human subject while simultaneously retaining the necessary and productive category of the human. This will allow us to theorize not just about the nature of the human, but also about the role of the human as moral agent, and allow also for a reflexive consideration of ourselves as part of a wider community of Earth-bound, finite beings. More pragmatically, it allows for an analysis of human relationships with technologies, which are neither utopian nor dystopian, but pragmatic.

But on what intellectual heritage should such a figure be built? Drawing boundaries is an inherently political act, contributing to what counts as human, how society is organized and its future potentiality. As Ruha Benjamin (2019: 9) argues for the project of robotics, 'the larger critical project involves questioning how robot technologies revolve around the category of "human" – "who defines it, inherits it, wields it ... who rents it, tills it, toils for it ... who gets expelled from it, buried under it, or drowned as they risk everything to inhabit it"'. Any reconfiguration of the concept of the human must include those who have previously been excluded from this domain. This should also be extended to other embodied beings with whom we share kinship and a planet. An alternative starting point is to propose an ontological ethic which posits an asymmetry between humans and technical objects on the one hand, and a radical continuity between humans and other animals on the other.

Haraway's cyborg remains a powerful vision of a posthumanist humanity, united in its diversity and in kinship with other embodied beings. It is a radical alternative, both to the transhumanist image as well as to critical dystopian readings of technology. The cyborg has deconstructed the entrenched and dominant view of the human and revealed its boundaries to be permeable and open for reconfiguration. Through the figure of the cyborg, Haraway has reoriented feminist accounts to the positive potential of science and technology in order to blur and transform (oppressive) boundaries and identities and instead offer a utopian vision of a post-gender, post-essentialist world. However, while the boundary between humans and other living organisms is indeed thoroughly breached, that with technology is not. Technologies have transformed our physical and cultural landscapes; they have 'recrafted our bodies' by extending our reach in the world and how we interact with others; they have even more markedly affected our political, social and environmental realities. But on a non-symbolic level, our experiencing bodies continue to share a greater commonality with all other embodied, living beings, than with machines. However, for Haraway, the cyborg is just one figuration among many other possibilities and reconfigurations (Åsberg 2024).

In her book *Wild Dog Dreaming*, anthropologist and environmentalist Rose (2011) proposes an 'ecological existentialism'. Existentialism extends phenomenological philosophy by focusing on the specifics of personal existence and the human condition. It arose as a response to the end of metaphysical certainty in the world at the end of the Second World War. Existentialism represents a deep awareness of humanity's loneliness and precarity in a wider universe devoid of meaning. It recognizes that there is no cosmic certainty, essential essence or teleological end point to meaning or to being human. Instead of order and predictability, there is ambiguity and flux. Thus, while science and technology are a quest for order and for abstract, universal laws and predictability, existential phenomenology starts instead from the idea of embodied existence. As well as loneliness, there is freedom: without a predetermined destiny, humans can choose their own path, to develop 'what appear to be astonishingly open ways of being and becoming human' (Rose 2011: 43). Out of this arises an ethical stance. If there is no higher order or organizing principle, we must find meaning and purpose ourselves in the activity of existence.

For Rose, an 'ecological' existentialism is a solution to the existential condition. As she articulates, the loneliness we experience as humans in the world is of our own making, through our progressive attempts to separate ourselves, to control and to dominate. We are, in fact, part of a much wider shared world and kinship with all life on Earth. Ecological existentialism thus reflects 'two major shifts in worldview: the end of certainty and the end of atomism' (Rose 2011: 2). This leads her to call for an acceptance of uncertainty and a focus on connectivity. Rose's ecological existentialism thus chimes with the ambiguity and kinship in Merleau-Ponty's phenomenological existentialism. Ultimately, for Rose: 'Ecological existentialism thus proposes a kinship of becoming: no telos, no deus ex machina to rescue us, no clockwork to keep us ticking along; and on the other hand, the rich plenitude, with all its joys and hazards, of our entanglement in the place, time, and multispecies complexities of life on Earth' (2011: 44). Rose (2011: 10) challenges us to find our way back 'home', to 'recover and discover' connectivity. In doing this we confront the suffering, death and perilousness that are a part of the living world. Indeed, it is by inhabiting our shared vulnerability that we recognize ourselves as embedded, finite and interdependent beings, rather than striving for technological transcendence or succumbing to apocalyptic and incapacitating despair. This kinship of becoming refigures the human as part of the flesh of planetary, embodied and affective life.

I thus propose a generative framework that brings the cyborg figure and the eco-existential human of Merleau-Ponty and Rose, though seemingly conflicting, into dialectical relation. In this way, rather than seeking synthesis or resolution, each can serve to both limit and illuminate one another. While the anti-essentialist, ontologically ironic human–machine hybrid cyborg frees us to reconfigure boundaries and celebrate hybridity and diversity, the eco-existential human reminds us that this freedom is rooted in our shared finitude and vulnerability with the living world.

CONCLUSION

What Robots Can Teach Us about Being Human in a Calamitous World

How hard we are willing to work for the future depends largely upon our image of what that future will be like. If we take the pessimistic view that human nature is getting progressively worse and our future will be grim, it is tempting to just give up, refuse to bring more children into the world, and to live out our lives consuming all the gasoline we can. If, on the other hand, we feel that it is possible to master our present-day problems, we can summon up the dedication and political will necessary to create a better world.

Margaret Mead (2005), *The World Ahead*

This book has explored the past and current work that is being done in AI and robotics research and its implications for how the human is known. Through deep ethnographic engagement with the field, in four robotics labs in three countries, I investigated what robots could teach us about being human. As well as investigating the spaces of robotics research, I also drew on extensive historical and philosophical material to reframe contemporary robotics research within the long history of human attempts to construct automata. This sees robots as a continuation of centuries of efforts by philosophers, mathematicians and engineers to reduce human activity to mathematical calculation. Throughout history, the human and the machine have served as models for one another, with conceptions of the human invariably mirroring the available scientific theories of the day.

Today, the view of the human as amenable to scientific and quantitative investigation continues to be upheld by a network of interrelated institutional orderings, including investors and venture capital, tech companies, governments, academic institutions and publishers.

Through these mechanisms, the human is conceived as scientifically knowable, despite a lack of scientific evidence. An extension of this incorrect assumption is the narrative that human-like intelligence, or even super-human intelligence, is imminent. This narrative is actively constructed by techno-utopian visions and a network of institutions reproducing the ideological, post-Enlightenment fantasies and visions of disembodied intelligence. Further, this image is being used to obscure the limitations of current technologies and the real work that is necessary to make the technologies appear intelligent and autonomous. This exaggerated image is also used to reorganize employment structures and practices, often to the detriment of workers.

Through first-hand engagement with robotics research, I uncovered the reality of AI and robotics work, helping to demystify what can seem as an opaque and impenetrable field. What I found is that by building and modifying robots, AI and robotics researchers are in many ways reflecting, reproducing and producing some of our foundational assumptions about what it means to be human. This includes the dualistic assumption of the mechanical body containing an information-processing mind, within an overarching evolutionary narrative.

However, robotics research is also interrupting these inherited assumptions. On entering the robotics lab, it quickly becomes apparent that the images shown in press releases and carefully managed demonstrations have more in common with science fiction than with reality. Through their research and engagement with the material realities of their craft, roboticists are forced into confrontation with the real. By engaging with the intricacies and nuances of trying to get a robot to work outside of the narrow confines of laboratories, roboticists are keenly aware of the limitations of current technologies. As a result, none believed AGI to be imminent. Thus, in stark contrast with dominant narratives, current developments in robotics and AI are in fact disproving two thousand years of philosophical theorizing that maintains that human, or non-human animal, intelligence is replicable in a machine.

Despite this, roboticists broadly subscribe to a materialistic and scientific view of the world, in which humans are ultimately knowable and reproducible, and thus all but one of my informants considered the achievement of human-like or super-human intelligence theoretically possible. Additionally, by carefully crafting and staging robot performances, as well as making the exaggerated claims required for funding proposals, robotics researchers are implicated in supporting these fantastical narratives in various ways.

Explicit and implicit theories of the human emerged in my research. These include that of the universal subject, whose rationality and 'intelligence' is the blueprint for AGI or even super-human intelligence. There is also the robot user, who is modelled in terms of 'mental models' and 'anthropomorphic projections' of psychology, often not 'fully' human, and subject to the determinants of evolution and biology. Less explicitly, there is also the invisible worker, whose gendered and racialized work is devalued and hidden, classed as unskilled and demeaning.

However, detailed observations of people's encounters with robots revealed an altogether different image of the human, one that challenges both the Enlightenment model of Man, as well as the symmetrical view of anthropology and STS theory. This is revealed in the many different ways in which humans and machines are entangled. In robot encounters, moving machines evoke a visceral and pre-discursive empathy in those that encounter them, regardless of their inner computational abilities. In their embodied interactions with robots, roboticists are engaged in creative, sensuous, skilful engagement with their materials and their craft, developing deep connections with their robots and a love for the technology. When performing the robot, what is revealed is people's immense spontaneity, skill and creativity, as well as their compassion and tenderness. In contrast to dominant narratives of competition and 'selfish genes', people feel a primal kinship with other 'bodies' in the world; it is in these moments that it becomes much more than just a technology or a tool.

These findings reflect those of previous laboratory studies which reveal how the boundary between the human and the machine is permeable and open for reconfiguration. However, there is a key difference. In all cases, it is the living body that animates the relation. In contrast, when not being performed, instead of being 'surprisingly lively', machines remain lifeless and inert. Humans and objects are indeed inextricably entangled and co-constituted, but they are not symmetrical.

What is revealed in the encounter is thus not symmetry, in which boundaries disappear and differences are negated; rather, what emerges is a palpable *asymmetry* between the human and the technical object. This contrasts strongly with studies of human–animal encounters, in which it is immediately and intuitively obvious to researchers that animals are not objects (see Huopalainen and Jammaers 2024; Kortetmäki 2019). It also contrasts with emerging research from across scientific and philosophical fields revealing the capacities of non-human animals for what were previously considered uniquely 'human' qualities, such as empathy, affect, care, sentience, vulnerability and self-affection. Thus, while there is a clear asymmetry between humans and machines, there is a radical continuity and kinship between humans and other living beings.

While symmetrical posthumanism extends critiques of the rational actor at the centre of economic theorizing, and offers intriguing opportunities to rethink and reconfigure relations, I argue that it is nonetheless limited by theorizing an equivalence between humans and technical objects. This limits both its utility in theorizing human complexity and situatedness, as well as its effectiveness in critiquing technical objects and the increasingly hegemonic power of technology companies.

In my research, the living body emerges as both a core category and an enduring material reality that cannot be reasoned away. Deviating from symmetrical ontology and the posthuman dissolution of the human, this work proposes an alternative cyborg figure, one that is continuous with the natural world but, in our living, feeling and creative bodies, radically discontinuous with human-made tools and technologies. In order to support this move, I have proposed a novel ontological ethic that brings the posthumanist cyborg figure into productive tension with the eco-existential human of embodied phenomenology and ecological existentialism. This builds on a posthumanist grounding to problematize and destabilize inherited boundaries and divisions, while also recognizing a reconfigured human in a state of continual becoming, rooted in finitude, vulnerability, intercorporeality and ecological kinship. If the cyborg reveals the contingency of categories, the eco-existential human reminds us of the stakes of embodiment, suffering and care in a shared world.

The implications of this new orientation will depend on the ways in which it is taken up and actualized. It is a useful counter to the myth that artificial human-level intelligence has been achieved or is imminent. That is not to say that AI is not powerful, useful or dangerous, but that the narrative of human-like intelligence is used to mystify and misapply it: artificially inflating expectations and investment in it, and concentrating and shifting power to those who own it and wield it. For researchers in both the fields of robotics and HRI, it is a call to engage more critically and reflexively in their work, recognizing it as political and world building. It also has implications for creative reimaginings in these fields, inviting roboticists to ask what it would mean to redesign technology, not in the image of a human but, rather, to reflect varied forms of organization and engagement with the world. It asks, what would it mean to cultivate a technic of care and finitude, rather than mastery and acceleration? For anthropologists, this reorientation is an invitation not just to deconstruct how boundaries are produced and reproduced in practice, but to engage in the positive work of constructing new ones. In particular, this should be oriented towards novel conceptual resources to address contemporary economic, political and ecological challenges. This new orientation asks what acknowledging the shared finitude and affinities between living, embodied beings can attune us to within wider cultural, political and ecological contexts. For researchers of all disciplines, it is also a call for methodological pluralism. As we have seen, robotics researchers find the requirements of their field growing, absorbing all domains. However, despite an openness in the field, they have limited grasp of, or appreciation for, alternative forms of knowledge that do not map easily to the engineering paradigm. A scientific investigation of the world needs to sit alongside an understanding of the world from a social and cultural perspective, rather than subsuming each other. Methodologically, it reminds us that alongside models and abstraction, certain empirical and ethical positions can be understood only in the concrete, face-to-face encounter. Thus, it is not just about embracing diverse forms of knowledge, but ensuring, as much as possible, that people themselves, in all their embodied wisdom and diversity, are present and empowered in decision making. While it is true that robotics and other technologies could potentially be enlisted to projects that promote greater equality, transformations and inclusion, robotics alone should not lead this reconfiguration.

For policy makers, the implication of a reconfigured figure of the human is an alternative to the image of the human in dominant economic theories. It also cautions against an overreliance on abstract, numerical, population-level data. Instead, it is a call to ensure that decision making and policy design are balanced with rich qualitative insights. It is also a call to expand the field of concern to all humans, and then also to other living beings with whom we share the planet and its finite resources.

If the only future we can imagine for ourselves is one of climate catastrophe, wars, inequity, degradation and exploitation, then it is tempting to just give up. If the choice is indeed between a calamitous Earth or conquering the universe through technology, then the latter is infinitely preferable. There is a deep desire, even fundamental need, as Joshua Schuster and Derek Woods (2021: 21) identify, for a 'secular eschatology', a vision of the future, for stories about the origins and goals of humanity, life and the universe.

Without that, there is little reason to work together for a common goal. However, when it comes visions of the future, this narrative has been ceded to the transhumanists.

Drawing a boundary between vulnerable life and ecological systems on the one hand, and technical and commodified objects on the other, interrupts this narrative. It is also a necessary precondition for ensuring clear accountability for technical harms, while providing political and regulatory clarity. Simultaneously, recognizing the vulnerabilities and interdependencies between humans and the rest of the living world allows us to imagine a world that is not based on exploitation and domination but on compassion and co-existence. Rather than eliminating the human as a significant conceptual category, this will involve reconfiguring the human as a cultural, embodied, situated and sensuous being, who is continuous, and kin, with all life on Earth.

References

Acemoglu, Daron, and Pascual Restrepo. 2017. 'Robots and Jobs: Evidence from US Labor Markets', *NBER Working Paper No. 23285*.

Adam, Alison. 1998. *Artificial Knowing: Gender and the Thinking Machine*. London: Routledge.

Adams, Douglas. 1979. *The Hitch Hiker's Guide to the Galaxy*. London: Pan Macmillan.

Agre, Philip E. 1997. *Computation and Human Experience*. Cambridge: Cambridge University Press.

Agre, Philip, and David Chapman. 1987. 'Pengi: An Implementation of a Theory of Activity', *Proceedings of the Sixth National Conference on Artificial Intelligence* 278: 268–72.

Alaimo, Stacy. 2010. *Bodily Natures: Science, Environment, and the Material Self*. Bloomington: Indiana University Press.

Altman, Sam. 2025. 'Reflections', *Sam Altman*. Retrieved 18 March 2025 from https://blog.samaltman.com/reflections.

American Society for Cybernetics. 2021. 'Summary: The Macy Conferences', *American Society for Cybernetics*. Retrieved 27 July 2021 from https://asc-cybernetics.org/foundations/history/MacySummary.htm.

——. 2025. 'ASC: Foundations: History: Timeline'. Retrieved 25 March 2025 from https://asc-cybernetics.org/foundations/timeline.htm.

Andes, Scott et al. 2017. 'Capturing the Next Economy: Pittsburgh's Rise as a Global Innovation City'. *Brookings Bass Initiative on Innovation and Placemaking*: 75.

Aristotle. 2000. *Politics* (B. Jowett, trans.) Mineola: Dover Publications.

Armstrong, Rachel. 2016. 'Transitioning towards the Ecocene', in Barbara Imhof and Petra Gruber (eds), *Built to Grow – Blending Architecture and Biology*. Switzerland: Birkhäuser Verlag, pp. 11–13.

Åsberg, Cecilia. 2024. 'Promises of Cyborgs: Feminist Practices of Posthumanities (Against the Nested Crises of the Anthropocene)', *Nordic Journal of Feminist and Gender Research* 32(2): 125–45.

Asimov, Isaac. 1942. 'Runaround', *Astounding Science Fiction*.

Atanasoski, Neda, and Kalindi Vora. 2019. *Surrogate Humanity: Race, Robots, and the Politics of Technological Futures*. Durham: Duke University Press.

Babbage, Charles. 1832. *On the Economy of Machinery and Manufactures*, 3rd edn. London: Charles Knight.

Baron-Cohen, Simon. 1991. 'Precursors to a Theory of Mind: Understanding Attention in Others', *Natural Theories of Mind: Evolution, Development and Simulation of Everyday Mindreading* 1: 233–51.

Bender, Shawn. 2025. *Feeling Machines: Japanese Robotics and the Global Entanglements of More-than-Human Care*. Stanford, California: Stanford University Press.

Benjamin, Ruha. 2019. 'Introduction: Discriminatory Design, Liberating Imagination', in Ruha Benjamin (ed.), *Captivating Technology: Race, Carceral Technoscience, and Liberatory Imagination in Everyday Life*. Durham: Duke University Press, pp. 1–22.

Boden, Margaret A. 2004. *The Creative Mind: Myths and Mechanisms, Second Edition*. London: Routledge.

———. 2006. 'Robots and Anthropomorphism', *AAAI Workshop Technical Report WS-06-09*: 69–74.

———. 2017. 'Principles of Robotics: Regulating Robots in the Real World', *Connection Science* 29(2): 124–29.

Bostrom, Nick. 2002. 'Existential Risks: Analyzing Human Extinction Scenarios and Related Hazards', *Journal of Evolution and Technology* 9: 1–30.

———. 2003. 'Ethical Issues in Advanced Artificial Intelligence', Nickbostrom.com. Retrieved 31 March 2025 from https://nickbostrom.com/ethics/ai.

———. 2005. 'A History of Transhumanist Thought', *Journal of Evolution and Technology* 14(1): 30.

———. 2013. 'Existential Risk Prevention as Global Priority', *Global Policy* 4(1): 15–31.

———. 2023. 'Apology for an Old Email'. Retrieved 8 January 2026 from nickbostrom.com/oldemail.pdf.

Boyer, George R. 1998. 'The Historical Background of the Communist Manifesto', *Journal of Economic Perspectives* 12(4): 151–74.

Boyer, Pascal. 1996. 'What Makes Anthropomorphism Natural: Intuitive Ontology and Cultural Representations', *The Journal of the Royal Anthropological Institute* 2(1): 83.

Breazeal, Cynthia L. 2002. *Designing Sociable Robots*. Cambridge, MA: The MIT Press.

Brooks, Rodney. 1990. 'Elephants Don't Play Chess', *Robotics and Autonomous Systems* 6(1–2): 3–15.

———. 1991a. 'Intelligence Without Reason'. A.I. Memo No. 1293. Retrieved 8 January 2026 from https://people.csail.mit.edu/brooks/papers/AIM-1293.pdf.

———. 1991b. 'Intelligence Without Representation', *Artificial Intelligence* 47: 139–59.

———. 1999. 'The Cog Project: Building a Humanoid Robot', in Chrystopher L. Nehaniv (ed.), *Computation for Metaphors, Analogy, and Agents*. Berlin, Heidelberg: Springer, pp. 52–87.

———. 2002. *Flesh and Machines: How Robots Will Change Us*. New York: Vintage Books.

Browning, Heather. 2023. 'Welfare Comparisons within and across Species', *Philosophical Studies* 180(2): 529–51.

Bryson, Joanna J. 2010. 'Robots Should Be Slaves', in Yorick Wilks (ed.), *Natural Language Processing* (Vol. 8). Amsterdam: John Benjamins Publishing Company, pp. 63–74.

Burke, J.L. et al. 2004. 'Final Report for the DARPA/NSF Interdisciplinary Study on Human–Robot Interaction', *IEEE Transactions on Systems, Man and Cybernetics, Part C (Applications and Reviews)* 34(2): 103–12.

Callon, Michel, and Bruno Latour. 1992. 'Don't Throw the Baby Out with the Bath School! A Reply to Collins and Yearley', in Andrew Pickering (ed.), *Science as Practice and Culture*. Chicago: University of Chicago Press, pp. 343–68.

Carpenter, Julie. 2013. *The Quiet Professional: An Investigation of U.S. Military Explosive Ordnance Disposal Personnel Interactions with Everyday Field Robots*. Thesis (PhD), University of Washington.

Castañeda, Claudia, and Lucy Suchman. 2014. 'Robot Visions', *Social Studies of Science* 44(3): 315–41.

CHAI. 2025. 'About – Center for Human-Compatible Artificial Intelligence'. Retrieved 26 March 2025 from https://humancompatible.ai/about/.

Chasin, Alexandra. 1995. 'Class and Its Close Relations: Identities among Women, Servants, and Machines', in Judith Halberstam and Ira Livingston (eds), *Posthuman Bodies, Unnatural Acts*. Bloomington: Indiana University Press, pp. 73–96.

Clynes, Manfred E., and Nathan S. Kline. 1960. 'Cyborgs and Space', *Astronautics* 14(9): 26–27.

Coalition for Critical Technology. 2021. 'Abolish the #TechToPrisonPipeline', *Medium*. Retrieved 22 November 2021 from https://medium.com/@CoalitionForCriticalTechnology/abolish-the-techtoprisonpipeline-9b5b14366b16.

Coeckelbergh, Mark. 2018. 'Why Care About Robots? Empathy, Moral Standing, and the Language of Suffering', *Kairos: Journal of Philosophy & Science* 20(1): 141–58.

Conty, Arianne Françoise. 2018. 'The Politics of Nature: New Materialist Responses to the Anthropocene', *Theory, Culture & Society* 35(7–8): 73–96.

Cowan, Ruth Schwartz. 1983. *More Work for Mother: The Ironies of Household Technology from the Open Hearth to the Microwave*. New York: Basic Books.

Crevier, Daniel. 1993. *AI: The Tumultuous History of the Search for Artificial Intelligence*. New York: Basic Books.

CSAIL. 2025. 'Cog Project Overview', *MIT Computer Science & Artificial Intelligence Laboratory*. Retrieved 25 March 2025 from https://groups.csail.mit.edu/lbr/humanoid-robotics-group/cog/overview.html.

CSER. 2024. 'Our Story – CSER'. Centre for the Study of Existential Risk website. Retrieved 26 March 2025 from https://www.cser.ac.uk/about-us/our-story/.

Cushing, Ellen. 2012. 'Amazon Mechanical Turk: The Digital Sweatshop', *Utne. Com* Retrieved 10 August 2021 from https://www.utne.com/science-and-technology/amazon-mechanical-turk-zm0z13jfzlin/.

Darling, Kate. 2016. 'Extending Legal Protection to Social Robots: The Effects of Anthropomorphism, Empathy, and Violent Behavior towards Robotic Objects', *Robot Law*. Cheltenham: Edward Elgar Publishing, pp. 213–32.

——. 2017. '"Who's Johnny?" Anthropomorphic Framing in Human–Robot Interaction, Integration, and Policy', in Patrick Lin et al. (eds), *Robot Ethics 2.0*. Oxford: Oxford University Press, pp. 173–92.

DARPA. [2025]. 'Shakey the Robot'. DARPA website. Retrieved 25 March 2025 from https://www.darpa.mil/node/2097.

Degeler, Andrii. 2015. 'Marines' LS3 Robotic Mule Is Too Loud for Real-World Combat', Ars Technica. Retrieved 10 November 2021 from https://arstechnica.com/information-technology/2015/12/us-militarys-ls3-robotic-mule-deemed-too-loud-for-real-world-combat/.

Dennett, D.C. 1998. *Kinds of Minds: Toward an Understanding of Consciousness*. New York: Basic Books.

Dreyfus, Hubert L. 1965. *Alchemy and Artificial Intelligence*. California: Rand Corporation.

——. 1972. *What Computers Can't Do; a Critique of Artificial Reason*. New York: Harper & Row.

——. 1992. *What Computers Still Can't Do: A Critique of Artificial Reason*. Cambridge, MA: MIT Press.

——. 2007. 'Why Heideggerian AI Failed and How Fixing It Would Require Making It More Heideggerian', *Philosophical Psychology* 20(2): 247–68.

Dreyfus, Hubert L., and Stuart Dreyfus. 2005. 'Peripheral Vision: Expertise in Real World Contexts', *Organization Studies* 26(5): 779–92.

——. 2009. *Mind over Machine: The Power of Human Intuition and Expertise in the Era of the Computer*. New York: The Free Press.

Duffy, Brian R. 2003. 'Anthropomorphism and the Social Robot', *Robotics and Autonomous Systems* 42(3–4): 177–90.

Ekbia, Hamid, and Bonnie Nardi. 2014. 'Heteromation and Its (Dis)Contents: The Invisible Division of Labor between Humans and Machines', *First Monday*.

Elish, M.C., and danah boyd. 2018. 'Situating Methods in the Magic of Big Data and AI', *Communication Monographs* 85(1): 57–80.

Eubanks, Virginia. 2019. *Automating Inequality: How High-Tech Tools Profile, Police, and Punish the Poor*, 1st edn. New York: Picador St. Martin's Press.

FHI. 2012. 'Future of Humanity Institute'. Retrieved 13 November 2021 from http://www.fhi.ox.ac.uk/.

FLI. 2025. 'Our Mission'. Future of Life Institute website. Retrieved 26 March 2025 from https://futureoflife.org/our-mission/.

Foucault, Michel. 2010. *The Birth of Biopolitics: Lectures at the Collège de France, 1978–79* (M. Senellart, ed.) Basingstoke: Palgrave Macmillan.

Fowles, Severin. 2016. 'The Perfect Subject (Postcolonial Object Studies)', *Journal of Material Culture* 21(1): 9–27.

Gallagher, Shaun, and Dan Zahavi. 2010. *The Phenomenological Mind: An Introduction to Philosophy of Mind and Cognitive Science*. London: Routledge.

Garreau, Joel. 2007. 'Bots on The Ground'. *Washington Post*, 5 May.

Gebru, Timnit, and Émile P. Torres. 2024. 'The TESCREAL Bundle: Eugenics and the Promise of Utopia through Artificial General Intelligence', *First Monday* 29(4).

Gill, Martha. 2019. 'Novacene by James Lovelock Review – Don't Worry, the Robots Might Not Kill Us', *The Times*, 21 June.

Grace, Katja et al. 2018. 'When Will AI Exceed Human Performance? Evidence from AI Experts'. Retrieved May 2024 from http://arxiv.org/abs/1705.08807.

——. 2024. 'Thousands of AI Authors on the Future of AI'. Retrieved 8 January 2026 from https://arxiv.org/abs/2401.02843.

Gray, Mary L., and Siddharth Suri. 2019. *Ghost Work: How to Stop Silicon Valley from Building a New Global Underclass*. New York: Houghton Mifflin Harcourt.

Grosz, E.A. 1994. *Volatile Bodies: Toward a Corporeal Feminism*. Bloomington: Indiana University Press.

Gunkel, David J. 2018. 'The Other Question: Can and Should Robots Have Rights?', *Ethics and Information Technology* 20(2): 87–99.

Guthrie, Stewart. 1993. *Faces in the Clouds: A New Theory of Religion*. New York: Oxford University Press.

Gutkind, Lee. 2009. *Almost Human: Making Robots Think*. London: W.W. Norton & Company.

Hao, Karen, and Deepa Seetharaman. 2023. 'Cleaning Up ChatGPT Takes Heavy Toll on Human Workers', *Wall Street Journal*, 24 July.

Haraway, Donna. 1987. 'A Manifesto for Cyborgs: Science, Technology, and Socialist Feminism in the 1980s', *Australian Feminist Studies* 2(4): 1–42.

——. 1989. *Primate Visions: Gender, Race and Nature in the World of Modern Science*. London: Routledge.

——. 1991. *Simians, Cyborgs, and Women: The Reinvention of Nature*. New York: Routledge.

——. 2016. *Staying with the Trouble: Making Kin in the Chthulucene*. Durham: Duke University Press.

Hayles, N. Katherine. 1999. *How We Became Posthuman: Virtual Bodies in Cybernetics, Literature, and Informatics*. Chicago: University of Chicago Press.

Hays, Jeffrey. 2013. 'Robots in Japan: History, Industrial Uses and Security', *Facts and Details*.

Heider, Fritz, and Marianne Simmel. 1944. 'An Experimental Study of Apparent Behavior', *The American Journal of Psychology* 57(2): 243.

Helmreich, Stefan. 2000. *Silicon Second Nature: Culturing Artificial Life in a Digital World* (Updated edition with a new preface, first paperback printing). London: University of California Press.

Henare, Amiria. 2003. 'Artefacts in Theory: Anthropology and Material Culture', *The Cambridge Journal of Anthropology* 23(2): 54–66.

Hirose, Masato, and Kenichi Ogawa. 2007. 'Honda Humanoid Robots Development', *Philosophical Transactions of the Royal Society A: Mathematical, Physical and Engineering Sciences* 365(1850): 11–19.

Howard, Christopher A., and Wendelin M. Küpers. 2020. 'Posthumanism and Anthropology', in Stefan Herbrechter et al. (eds), *Palgrave Handbook of Critical Posthumanism*. London: Palgrave Macmillan, pp. 1–23.

Huopalainen, Astrid, and Eline Jammaers. 2024. '"Making the Unimaginable Imaginable?" The Power of Artmaking in Understanding Animal Vulnerabilities and "Humanimal" Relationality in Organization Studies', *Organization* 32(6): 841–60.

Huyssen, Andreas. 1986. *After the Great Divide: Modernism, Mass Culture, Postmodernism*. Bloomington: Indiana University Press.

IEEE Standards Association. 2016. *Safety and Beneficence of Artificial General Intelligence (AGI) 4 and Artificial Superintelligence (ASI)*. IEEE.

Ieropoulos, Ioannis A. et al. 2013. 'Waste to Real Energy: The First MFC Powered Mobile Phone', *Physical Chemistry Chemical Physics* 15(37): 15312.

Ingold, Tim. 1997. 'Eight Themes in the Anthropology of Technology', *Social Analysis: The International Journal of Anthropology* 41(1): 106–38.

——. 2000. *The Perception of the Environment*. London: Routledge.

Jackson, Michael. 2002. 'Familiar and Foreign Bodies: A Phenomenological Exploration of the Human–Technology Interface', *The Journal of the Royal Anthropological Institute* 8(2): 333–46.

Jasanoff, Sheila, and Sang-Hyun Kim (eds). 2015. *Dreamscapes of Modernity: Sociotechnical Imaginaries and the Fabrication of Power*. Chicago: University of Chicago Press.

Jin, Berber, and Joanna Stern. 2025. 'Anthropic CEO Says AI Could Surpass Human Intelligence by 2027', *Wall Street Journal*, 21 January.

Kaempffert, Waldemar. 1927. 'Science Produces the "Electrical Man"', *New York Times*.

Katz, Yarden. 2017. 'Manufacturing an Artificial Intelligence Revolution'. *SSRN preprint: 3078224*. Retrieved March 2025 from https://papers.ssrn.com/abstract=3078224.

Kember, Sarah. 2003. *Cyberfeminism and Artificial Life*. London and New York: Routledge.

Kim, Kyungtae. 2025. 'What Is AGI, and When Will It Arrive? Big Tech CEO Predictions', *GizAI*. Retrieved 20 March 2025 from https://www.giz.ai/what-is-agi-and-when-will-it-arrive/.

King, Elizabeth. 2007. 'Perpetual Devotion: A Sixteenth-Century Machine', in Jessica Riskin (ed.), *Genesis Redux: Essays in the History and Philosophy of Artificial Life*. Chicago: University of Chicago Press, pp. 263–92.

Knorr-Cetina, K.D. 2014. *The Manufacture of Knowledge: An Essay on the Constructivist and Contextual Nature of Science*. Kent: Elsevier Science.

Kodate, Naonori et al. 2022. 'Hopes and Fears Regarding Care Robots: Content Analysis of Newspapers in East Asia and Western Europe, 2001–2020', *Frontiers in Rehabilitation Sciences* 3.

——. 2024. 'Assembling Sociality in Caring Spaces: Culturally Sensitive Robot Deployment During the Pandemic in Residential Care Homes in Ireland and Japan', *Engaging Science, Technology, and Society* 10(3): 181–206.

Kolb, Michael. 2012. *Soldier and Robot Interaction in Combat Environments*. Thesis (PhD), University of Oklahoma.

Kortetmäki, Teea. 2019. 'The Reification of Non-Human Nature', *Environmental Values* 28(4): 489–506.

Kuipers, Benjamin et al. 2017. 'Shakey: From Conception to History', *AI Magazine* 38(1): 88–103.

Küpers, Wendelin M. 2020. 'From the Anthropocene to an "Ecocene"? Eco-Phenomenological Perspectives on Embodied, Anthrodecentric Transformations towards Enlivening Practices of Organising Sustainably', *Sustainability* 12(9): 3633.

Kurzweil, Ray. 2005. *The Singularity Is Near: When Humans Transcend Biology*. New York: Viking.

——. 2014. *The Age of Spiritual Machines: When Computers Exceed Human Intelligence*. New York: Penguin Books.

Latour, Bruno. 1983. 'Give Me a Laboratory and I Will Raise the World', in Karin Knorr-Cetina and Michael Mulkay (eds), *Science Observed: Perspectives on the Social Study of Science*. London: Sage, pp. 141–70.

——. 1987. *Science in Action: How to Follow Scientists and Engineers through Society*. Cambridge, MA: Harvard University Press.

——. 1993. *We Have Never Been Modern*. Cambridge, MA: Harvard University Press.

——. 2002. 'Morality and Technology: The End of the Means', *Theory, Culture & Society* 19(5–6): 247–60.

——. 2004. *Politics of Nature: How to Bring the Sciences into Democracy*. Cambridge, MA: Harvard University Press.

——. 2014. 'Agency at the Time of the Anthropocene', *New Literary History* 45(1): 1–18.

——. 2015. 'Telling Friends from Foes in the Time of the Anthropocene', in Clive Hamilton et al. (eds), *The Anthropocene and the Global Environmental Crisis; Rethinking Modernity in a New Epoch*. London and New York: Taylor & Francis, pp. 145–55.

——. 2017a. 'Why Gaia Is Not a God of Totality', *Theory, Culture & Society* 34(2–3): 61–81.

——. 2017b. 'Anthropology at the Time of the Anthropocene: A Personal View of What Is to Be Studied', in Marc Brightman and Jerome Lewis (eds), *The Anthropology of Sustainability*. New York: Palgrave Macmillan, pp. 35–49.

Latour, Bruno, and Steve Woolgar. 1986. *Laboratory Life: The Construction of Scientific Facts*. Princeton, NJ: Princeton University Press.

LeCun, Yann et al. 2015. 'Deep Learning', *Nature* 521(7553): 436–44.

Lefebvre, Henri. 1991. *The Production of Space* (D. Nicholson-Smith, trans.) Malden: Blackwell Publishing.

Leopold, Todd. 2015. 'HitchBOT, Hitchhiking Robot, Gets Beheaded in Philly', *CNN*. Retrieved 11 August 2021 from https://edition.cnn.com/2015/08/03/us/hitchbot-robot-beheaded-philadelphia-feat/index.html.

Letizia, Anthony. 2013. 'The Robots of Westinghouse', *Geekfrontiers.Com*. Retrieved 27 July 2021 from https://geekfrontiers.com/pittsburgh-history/the-robots-of-westinghouse/.

Levinas, Emmanuel. 1999. *Otherwise Than Being, or, Beyond Essence* (A. Lingis, trans.) Pittsburgh, PA: Duquesne University Press.

Levi-Strauss, Claude. 1966. *The Savage Mind*. Chicago and London: University of Chicago Press.

Lighthill, James. 1973. *Lighthill Report: Artificial Intelligence: A Paper Symposium*. UK: Science Research Council.

Lukács, Georg. 2017. 'Reification and the Consciousness of the Proletariat', in Bertell Ollman and Kevin B. Anderson (eds), *Karl Marx* (1st edn). Abingdon: Routledge, pp. 3–25.

Maher, Kris. 2019. 'Pittsburgh Air-Quality Problem Recalls the Bad Old Days', *The Wall Street Journal*, 12 March.

Mak, Aaron. 2025. 'What's up with Peter Thiel and the Antichrist?', *POLITICO*, 24 October.

Marcus, Gary. 2018. 'Deep Learning: A Critical Appraisal'. Retrieved February 2026 from http://arxiv.org/abs/1801.00631.

——. 2022. 'Deep Learning Is Hitting a Wall', *Nautilus*, 10 March.

Marcus, George E. 1995. 'Ethnography in/of the World System: The Emergence of Multi-Sited Ethnography', *Annual Review of Anthropology* 24: 95–117.

Marsh, Allison. 2018. 'Elektro the Moto-Man Had the Biggest Brain at the 1939 World's Fair', *IEEE Spectrum* 6.

Marx, Paris. 2023. 'The ChatGPT Revolution Is Another Tech Fantasy', *Disconnect*, 27 July.

Masís, Jethro. 2014. 'Making AI Philosophical Again: On Philip E Agre's Legacy', *Continent* 4(1).

Mateescu, Alexandra, and Madeleine Clare Elish. 2019. *AI in Context: The Labour of Integrating New Technologies* (No. 30), Data & Society.

Matthews, Dylan. 2015. 'I Spent a Weekend at Google Talking with Nerds about Charity I Came Away ... Worried', *Vox*, 10 August.

McCarthy, John et al. 1955. *Dartmouth Summer Research Project on Artificial Intelligence*. Retrieved 8 January 2026 from http://jmc.stanford.edu/articles/dartmouth/dartmouth.pdf.

——. 1989. 'Artificial Intelligence, Logic and Formalizing Common Sense', in Richmond H. Thomason (ed.), *Philosophical Logic and Artificial Intelligence*. Dordrecht: Springer Netherlands, pp. 161–90.

McCorduck, Pamela. 2004. *Machines Who Think: A Personal Inquiry into the History and Prospects of Artificial Intelligence*. Massachusetts: A.K. Peters.
McCulloch, Warren, and Walter H. Pitts. 1943. 'A Logical Calculus of the Ideas Immanent in Nervous Activity', *The Bulletin of Mathematical Biophysics* 5: 115–33.
McFarland, Matt. 2014. 'Elon Musk: "With Artificial Intelligence We Are Summoning the Demon"', *Washington Post*, 24 October.
Mead, Margaret. 2005. *The World Ahead: An Anthropologist Anticipates the Future*. New York: Berghahn Books.
Mednick, Sam et al. 2025. 'How US Tech Giants Supplied Israel with AI Models, Raising Questions about Tech's Role in Warfare', *AP News*, 18 February.
Mellor, David J. 2019. 'Welfare-Aligned Sentience: Enhanced Capacities to Experience, Interact, Anticipate, Choose and Survive', *Animals: An Open Access Journal from MDPI* 9(7): 440.
Merleau-Ponty, Maurice. 1968. *The Visible and the Invisible: Followed by Working Notes* (C. Lefort and A. Lingis, eds), Studies in Phenomenology and Existential Philosophy. Evanston: Northwestern University Press.
——. 2012. *Phenomenology of Perception* (D.A. Landes, ed.) New York: Routledge.
Miller, Daniel. 1987. *Material Culture and Mass Consumption*. Oxford: Wiley Blackwell.
Minsky, Marvin, and Seymour Papert. 2017. *Perceptrons: An Introduction to Computational Geometry* (2017 edition). Cambridge, MA: The MIT Press.
Minsky, Marvin, and Doug Riecken. 1994. 'A Conversation with Marvin Minsky About Agents', *Communication of the ACM* 37(7): 22–29.
MIRI. 2024, 17 October. 'About the Machine Intelligence Research Institute', *MIRI*. Retrieved 26 March 2025 from https://intelligence.org/about/.
Moravec, Hans. 1995. *Mind Children: The Future of Robot and Human Intelligence*. Cambridge, MA: Harvard University Press.
——. 1998. 'When Will Computer Hardware Match the Human Brain?', *Journal of Evolution and Technology* 1(1): 10.
Mori, Masahiro. 1970. 'The Uncanny Valley', *Energy* 7: 33–35.
——. 2012. 'The Uncanny Valley: The Original Essay by Masahiro Mori' (K.F. MacDorman and N. Kageki, trans.), *IEEE Spectrum* 6.
Morris, Stephen, and Rafe Uddin. 2025. 'Big Tech Lines up over $300bn in AI Spending for 2025', *Financial Times*, 7 February.
Moya, Patricia. 2014. 'Habit and Embodiment in Merleau-Ponty', *Frontiers in Human Neuroscience 8*.
Naughton, John. 2021. 'Worried about Super-Intelligent Machines? They Are Already Here', *The Observer*, 25 December.
Nichols, Greg. 2021. 'Delivery Robots Are Taking over College Campuses', *ZDNet*.
Nilsson, Nils J. 2010. *The Quest for Artificial Intelligence: A History of Ideas and Achievements*. New York: Cambridge University Press.
Nomura, Tatsuya et al. 2015. 'Why Do Children Abuse Robots?', *Proceedings of the Tenth Annual ACM/IEEE International Conference on Human–Robot*

Interaction Extended Abstracts. HRI '15: ACM/IEEE International Conference on Human–Robot Interaction.

Norton-Wise, M. 2007. 'The Gender of Automata in Victorian Britain', in Jessica Riskin (ed.), *Genesis Redux: Essays in the History and Philosophy of Artificial Life*. Chicago: University of Chicago Press, pp. 163–295.

Nussbaum, Martha C. 2001. *The Fragility of Goodness: Luck and Ethics in Greek Tragedy and Philosophy*. New York: Cambridge University Press.

——. 2024. *Justice for Animals: Our Collective Responsibility*. (First Simon & Schuster trade paperback edition). New York: Simon & Schuster Paperbacks.

Ocasio-Cortez, Alexandria. 2021. 'Shout out to everyone who fought against community advocates who demanded ...' *@AOC*. Twitter, 25 February. https://twitter.com/AOC/status/1365021717144420354.

Olinerd. 2008, 17 March. 'Boston Dynamics Big Dog'. Retrieved 19 August 2021 from https://www.youtube.com/watch?v=W1czBcnX1Ww.

O'Neil, Cathy. 2017. *Weapons of Math Destruction: How Big Data Increases Inequality and Threatens Democracy*. London: Penguin Books.

OpenAI. 2025. 'OpenAI Charter'. Retrieved 21 March 2025 from https://openai.com/charter/.

Papert, Seymour. 1968. *The Artificial Intelligence of Hubert L. Dreyfus: A Budget of Fallacies* (MIT Project Mac Artificial Intelligence Memo. No. 154).

Pérez-Manrique, Ana, and Antoni Gomila. 2018. 'The Comparative Study of Empathy: Sympathetic Concern and Empathic Perspective-Taking in Non-Human Animals', *Biological Reviews* 93(1): 248–69.

Perrigo, Billy. 2023. 'Exclusive: The $2 Per Hour Workers Who Made ChatGPT Safer', *TIME*, 18 January.

Purtill, Corinne. 2019. 'The Robot That Could Change the Senior Care Industry: Time', *TIME*, 4 October.

Raworth, Kate. 2022. *Doughnut Economics: Seven Ways to Think like a 21st-Century Economist*. London: Penguin Books.

Richardson, Kathleen. 2015. *An Anthropology of Robots and AI: Annihilation Anxiety and Machines*, Routledge Studies in Anthropology. New York: Taylor & Francis.

Riskin, Jessica. 2003. 'The Defecating Duck, or, the Ambiguous Origins of Artificial Life', *Critical Inquiry* 29(4): 599–633.

—— (ed.) 2007. *Genesis Redux: Essays in the History and Philosophy of Artificial Life*. Chicago: University of Chicago Press.

Robertson, Jennifer. 2007. 'Robo Sapiens Japanicus: Humanoid Robots and the Posthuman Family', *Critical Asian Studies* 39(3): 369–98.

Rose, Deborah Bird. 2011. *Wild Dog Dreaming: Love and Extinction*, Under the Sign of Nature: Explorations in Ecocriticism. Charlottesville: University of Virginia Press.

Rosenblatt, Frank. 1957. *The Perceptron: A Perceiving and Recognizing Automaton (Project Para)* (Report No. 85-460-1). Buffalo: Cornell Aeronautical Laboratory, Inc.

Rosenblueth, Arturo et al. 1943. 'Behaviour, Purpose and Teleology', *Philosophy of Science* 10(1): 18–24.

Rosenthal-von der Pütten, Astrid M. et al. 2013. 'An Experimental Study on Emotional Reactions towards a Robot', *International Journal of Social Robotics* 5(1): 17–34.

Rumelhart, David E. et al. 1987. 'A General Framework for Parallel Distributed Processing', *Parallel Distributed Processing: Explorations in the Microstructure of Cognition: Foundations*. Parallel Distributed Processing: Explorations in the Microstructure of Cognition: Foundations, Cambridge, MA: MIT Press, pp. 45–76.

Russell, Stuart J., and Peter Norvig. 1995. *Artificial Intelligence: A Modern Approach*. Englewood Cliffs: Prentice Hall.

Saerbeck, Martin, and Christoph Bartneck. 2010. 'Perception of Affect Elicited by Robot Motion', *2010 5th ACM/IEEE International Conference on Human–Robot Interaction (HRI)*. 2010 5th ACM/IEEE International Conference on Human–Robot Interaction (HRI).

Sample, Ian. 2015. 'AI: Will the Machines Ever Rise Up?', *Guardian*, 26 June.

Scassellati, Brian. 2002. 'Theory of Mind for a Humanoid Robot', *Autonomous Robots* 12(1): 13–24.

Schuster, Joshua, and Derek Woods. 2021. *Calamity Theory: Three Critiques of Existential Risk*. Minneapolis: University of Minnesota Press.

Shannon, C.E. 1948. 'A Mathematical Theory of Communication', *The Bell System Technical Journal* 27: 379–423, 623–56.

Sharkey, Noel. 2007. 'The Programmable Robot of Ancient Greece', *New Scientist*, 4 July.

Shea, Michael. 2015. 'Karakuri: Subtle Trickery in Device Art and Robotics Demonstrations at Miraikan', *Leonardo* 48(1): 40–47.

Shelley, Mary. 1993. *Frankenstein, or, The Modern Prometheus*. Herts: Wordsworth Classics.

Shumailov, Ilia et al. 2024. 'The Curse of Recursion: Training on Generated Data Makes Models Forget'. *arXiv:2305.17493*.

Simon, Herbert A. 1996. *Models of My Life*, 1st edn. Cambridge, MA: MIT Press.

Simon, Herbert A., and Allen Newell. 1958. 'Heuristic Problem Solving: The Next Advance in Operations Research', *Operations Research* 6(1): 1–10.

Singer, P.W. 2010. *Wired for War: The Robotics Revolution and Conflict in the Twenty-First Century*. New York: Penguin Books.

Star Wars. 2014. 'The Cinema behind Star Wars: Metropolis', *StarWars.Com*. Retrieved 10 November 2021 from https://www.starwars.com/news/the-cinema-behind-star-wars-metropolis.

Starship Careers. 2019. 'Starship Technologies – Robot Operator on Campus'. Retrieved 30 September 2019 from https://jobs.lever.co/starship/6386c6ed-04fd-4ee4-8fae-d8e55768de12.

Statista. 2024. 'Data Growth Worldwide 2010–2028'. Statista website. Retrieved 6 March 2025 from https://www.statista.com/statistics/871513/worldwide-data-created/.

Suchman, Lucy. 1985. *Plans and Situated Actions: The Problem of Human–Machine Communication*. Palo Alto: Xerox Corporation.

——. 2007. *Human–Machine Reconfigurations: Plans and Situated Actions*, 2nd edn. Cambridge: Cambridge University Press.

——. 2011. 'Anthropological Relocations and the Limits of Design', *Annual Review of Anthropology* 40(1): 1–18.

——. 2019. 'Demystifying the Intelligent Machine', in Teresa Heffernan (ed.), *Cyborg Futures: Cross-disciplinary Perspectives on Artificial Intelligence and Robotics*. London: Palgrave Macmillan, pp. 35–61.

Suzuki, Yutaka et al. 2015. 'Measuring Empathy for Human and Robot Hand Pain Using Electroencephalography', *Scientific Reports* 5(1): 15924.

Takanishi, Atsuo. 2019. 'Historical Perspective of Humanoid Robot Research in Asia', in Ambarish Goswami and Prahlad Vadakkepat (eds), *Humanoid Robotics: A Reference*. Dordrecht: Springer Netherlands, pp. 35–52.

Tănăsescu, Mihnea. 2022. *Ecocene Politics*. Cambridge: Open Book Publishers.

Taussig, Michael. 1993. *Mimesis and Alterity: A Particular History of the Senses*. New York: Routledge.

Taussig, Michael T. 2010. *The Devil and Commodity Fetishism in South America* (30th anniversary edn). Chapel Hill: University of North Carolina Press.

Turing, A.M. 1937. 'On Computable Numbers, with an Application to the Entscheidungsproblem', *Proceedings of the London Mathematical Society* 2(1): 230–65.

Turkle, Sherry. 2011. *Alone Together: Why We Expect More from Technology and Less from Each Other*. New York: Basic books.

Uddin, Rafe, and Stephen Morris. 2025. 'Big Tech Lines up over $300bn in AI Spending for 2025', *Financial Times*, 7 February.

UNDP. 2017. 'UNDP in Asia and the Pacific Appoints World's First Non-Human Innovation Champion'. United Nations lead agency on international development in Asia and the Pacific. Retrieved 8 November 2021 from https://www.asia-pacific.undp.org/content/rbap/en/home/presscenter/pressreleases/2017/11/22/rbfsingapore.html.

Verdonck, Kris. 2016. 'DANCER #1'. *Atwodogscompany.* Retrieved 30 May 2016 from https://www.youtube.com/watch?v=xKbBImj3mWM.

Walter, William Grey. 1963. *The Living Brain*, 5th edn. New York: Norton.

Wehrle, Maren. 2021. '"Bodies (That) Matter": The Role of Habit Formation for Identity', *Phenomenology and the Cognitive Sciences* 20(2): 365–86.

Weizenbaum, Joseph. 1976. *Computer Power and Human Reason: From Judgment to Calculation*. San Francisco: W.H. Freeman.

Westfahl, Gary. 1992. '"The Jules Verne, HG Wells, and Edgar Allan Poe Type of Story": Hugo Gernsback's History of Science Fiction', *Science Fiction Studies* 19(3): 340–53.

Westling, Louise. 2010. 'Merleau-Ponty's Human-Animality Intertwining and the Animal Question', *Configurations* 18: 161–80.

Whelan, Christal. 2011. 'Robot Revolution: Significant Otherness in Japan', *The Daily Yomiuri*.
——. 2014. *Kansai Cool: A Journey into the Cultural Heartland of Japan*. North Clarendon, VT: Tuttle Publishing.
Wiener, Norbert. 1949. 'Cybernetics or Control and Communication in the Animal and the Machine', *Psychological Bulletin* 46(3): 236–37.
Winner, Langdon. 1980. 'Do Artifacts Have Politics?', *Daedalus* 109(1): 121–36.
World Economic Forum. 2011. 'Personal Data: The Emergence of a New Asset Class'. World Economic Forum, Bain & Company, Inc, pp. 1–40.
Wrage, Birte. 2022. 'Caring Animals and Care Ethics', *Biology & Philosophy* 37(3): 18.
Wright, James. 2023. *Robots Won't Save Japan: An Ethnography of Eldercare Automation*, Ithaca: ILR Press.
Zewe, Adam. 2025. 'Explained: Generative AI's Environmental Impact', *MIT News,* 17 January.

Index

www.ingramcontent.com/pod-product-compliance
Ingram Content Group UK Ltd.
Pitfield, Milton Keynes, MK11 3LW, UK
UKHW021836150926
8656IPUK00004B/17

9 781836 955887